Private Partnerships and Public Networks in Europe

Harry Cowie
Rapporteur

FEDERAL TRUST REPORT

THE FEDERAL TRUST

The Federal Trust works through research and education towards the widening and deepening of the European Union as well as to enhance the European policy of the United Kingdom

The Federal Trust conducts enquiries, promotes seminars and conferences and publishes reports on a wide range of contemporary issues. Its current work programme includes continuing study of the Intergovernmental Conference of 1996 and of the developing Economic and Monetary Union. The Trust runs projects on the building of alternative scenarios for the future of Europe and on the EU's potential contribution to employment.

The Trust also runs an education programme for sixth forms, universities and young leaders, and is leading EU-wide work on improving the European dimension in schools.

The Federal Trust is the UK member of TEPSA (the Trans-European Policy Studies Association).

The cover photograph is the Eurostar terminal at Waterloo Station, London © Peter Cook.

Published by the Federal Trust
11 Tufton Street
London SW1P 3QB

© Federal Trust for Education and Research, 1996

ISBN 0 90157371 X

THE FEDERAL TRUST IS A REGISTERED CHARITY

MARKETING AND DISTRIBUTION BY SWEET & MAXWELL LTD
PRINTED IN GREAT BRITAIN

Contents

Preface

The Federal Trust established a high-level study group to investigate the role and potential of public-private partnerships in the financing, delivery and operation of infrastructure investment in the European Union, particularly the Trans-European Networks (TENs). The inquiry, *Private Enterprise and Public Utility in the European Union*, began in January 1996 and examined best practice in Europe and elsewhere. The study group analysed the problems inherent in cross-border projects and in this report it makes policy recommendations with the aim of accelerating the development of public infrastructure on the basis mainly of private finance.

The terms of reference of the study group were:

- to analyse the problems involved in designing, building, financing and operating trans-European networks, with special reference to the fields of transport, energy and telecommunications;
- to examine best practice in developing public-private partnerships in Europe and elsewhere in all sectors of public utility;
- to consider the development of relevant policy instruments at the level of the European Union to promote competitiveness, efficiency, employment and environmental protection.

The membership of the study group was:

Tom Barrett	Head of Division, Transport and Energy, EIB
Sheila Beck	Marketing Manager, Siemens Business Services
Chris Boyd	Office of Neil Kinnock, European Commission
Nick Butler	Group Policy Adviser, BP
John Carr	Partner, Corporate Finance, Price Waterhouse
Sir Brian Corby	*Chairman*
Harry Cowie	Senior Research Fellow, Federal Trust, *Rapporteur*
Andrew Duff	Director, Federal Trust
Wolfgang Hager	Director, ECIS
Geoffrey Haley	Partner, S.J. Berwin

Members served in their personal capacity; they do not necessarily concur with all the views expressed in the Report, but they support its general conclusions, and welcome its publication.

The Federal Trust is an independent charity and as such holds no political view of its own.

The Trust is particularly grateful to Sir Brian Corby for his skilful chairmanship, and to Harry Cowie, who acted as secretary and who largely wrote the report. Martin Lodge and Richard Blackman helped with the research and organisation.

DG VII of the European Commission, the European Investment Bank and BP were generous sponsors of the project.

Comments on the Report would be welcome, and should be addressed to Harry Cowie.

Andrew Duff
Director

The Federal Trust
11 Tufton Street
London SW1P 3QB   September 1996

Foreword

It is common ground that in order to achieve the full potential of the single market in Europe significant improvements need to be made in its infrastructure. There are major missing links. Many telecommunications, transport and energy industries are still national state monopolies. This fragments the networks, restricts their vision and leaves them short of cash. Moreover, their monopolistic status discourages the private sector from providing the missing links.

The priority is to open up the provision of infrastructure to competition while recognising that there is an important element of public utility. Public-private partnerships can provide the way forward. They provide extra sources of finance while ensuring that investment responds to commercial needs rather than purely political pressures. Further, the involvement of the private sector brings new skills and know-how.

It would be naive, however, to think that liberalisation can solve all the problems. Progress so far in implementing the priority trans-European transport networks has been disappointingly slow. Many rail and road projects which are justified on a wider cost-benefit taking account of the European dimension are insufficiently profitable to be financed entirely by private sector support. Innovative ways of managing public and private finance are required, including public equity and electronic tolls.

This Federal Trust report examines the underlying reasons for the delays, including the problems for civil servants whether in Brussels or Whitehall in making entrepreneurial decisions especially when they have a public accounts committee looking over their shoulder. Our suggestion for the establishment of a European Infrastructure Agency is designed to ensure that the final political decisions about priorities and public funding are taken after an independent, expert examination of the issues which would be published and subject to public debate.

I commend the report.

Sir Brian Corby
Chairman of the Study Group

Glossary

CEE	Central and Eastern Europe
COURLY	Communauté urbaine de Lyon
CTRL	Channel Tunnel Rail Link
DBFO	Design-build-finance-operate
DCMF	Design-construct-manage-finance
DG	Directorate-General
EBRD	European Bank for Reconstruction and Development
EC	European Community
ECIS	European Centre for Infrastructure Studies
Ecofin	Council of Economic and Finance Ministers
Ecu	European currency unit
EEIG	European Economic Interest Grouping
EIB	European Investment Bank
EIF	European Investment Fund
EMU	Economic and Monetary Union
ERT	European Round Table of Industrialists
EU	European Union
HST	High-Speed Train
IGC	Intergovernmental Conference
IRR	Internal rate of return
LCR	London & Continental Railway
NEFCO	Nordic Environmental Finance Corporation
NERA	National Economic Research Associates
NPV	Net present value
OECD	Organisation for Economic Cooperation and Development
ONP	Open Network Provision
PBKAL	Paris-Brussels-Cologne-Amsterdam-London
PFI	Private Finance Initiative
Phare	Pologne et Hongrie: Aide à la Restructuration Economique

QMV	Qualified majority voting
SNCB	Société Nationale des Chemins de Fer Belges
SNCF	Société Nationale des Chemins de Fer Français
TEN	Trans-European Network
TGV	Train à Grande Vitesse
TML	Trans-Manche Link
UDAG	Urban Development Action Grant (US)
UDG	Urban Development Grant (UK)
WTO	World Trade Organisation

Summary

1. A major factor in improving the competitiveness of European business lies in the modernisation of its infrastructure in terms of transport, energy supply and telecommunications. But neither the quantity of public sector finance nor the quality of public authority management is sufficient to network Europe: the heavy involvement of the private sector is required at every stage of infrastructure development. A new methodology is required to achieve optimal partnerships between private enterprise and public utility.

2. The European Union's objective of establishing a system of priority Trans-European Networks as a major step to providing the missing links in European infrastructure is unlikely to be accomplished in the time-scale that was originally envisaged.

3. The difficulties facing the priority projects flow from one or more of these properties:

*   trans-frontier nature
*   large scale
*   high risk
*   limited financial viability
*   political will.

4. Member state governments are faced with major budgetary constraints in the run-up to the third stage of EMU and beyond. They are unwilling to provide the necessary grant aid or public finance to bridge the gap between what is an acceptable rate of financial and economic return to the private sector and the benefit to public utility measured in terms of the European Union as a whole.

5. The Delors White Paper of 1993 looked to public-private partnerships to provide both finance and expertise for the priority projects. Such partnerships pool technical expertise at the outset of a project, combine financial resources and share the risk. Risk

1

should be allocated between the partners according to where it can best be managed over the long term. Public-private partnerships transfer new entrepreneurial skills into the public sector. Their goal is to design, build, finance and operate the networks. Public-private partnerships will also facilitate the process of EU enlargement to Central and Eastern Europe, where public finance is in short supply.

6. The Essen European Council in 1994 chose twenty-four priority TENs. Those in the field of transport were all large-scale and ambitious, and some are proving problematical. High expectations that public-private partnerships would drive the TENs programme forward have not yet been realised. There are exciting prospects including the Channel Tunnel Rail Link, where existing revenues from Eurostar and substantial government grants of land have been made available. In other cases, political will appears to be faltering.

7. In the energy and telecoms sectors the role of private finance is already established and growing in importance as a result of deregulation, competition and privatisation. For transport there is still a shortage of finance. To minimise the risks of entering a new market, companies need to build up a portfolio of infrastructure schemes. European capital markets need to generate both more venture capital and equity for long-term investments. Grant-aid from the public sector is still important, but the public authorities would also stimulate infrastructure development by taking equity stakes in TENs. At the same time the new financial structures would attract pension and insurance fund investment.

8. The transport TENs are now in danger of stalling. The European Council in June 1996 did not endorse President Santer's proposals for additional EU grant financing for TENs. Instead it set up a High-Level Group to press for further progress. It is difficult to see how yet another expert group will be able to reconcile the problems involved without a secretariat and some increase in administrative expenditure.

9. The European Commission should set up immediately a new Task Force, with the powers of a directorate-general, to bring together the existing expertise which is at present dispersed among

its various directorates and other European institutions. The remit of the Task Force should be to expedite the TENs programme and to galvanise public-private partnerships.

10. The Task Force and the High-Level Group should together prepare the ground for the establishment of an independent high level body — a European Infrastructure Agency — with powers to evaluate and coordinate TENs projects from a European dimension. Such an agency would be able to draw on the expertise of the European Commission, the European Investment Bank, the European Investment Fund, the EBRD and member governments and corporations as well as the private sector.

11. In particular, the *European Infrastructure Agency* would:

- promote the use of best practice in evaluating projects thereby permitting a more rational prioritisation of projects;
- establish a methodology to ensure that the technical, social and economic analysis takes full account of the whole European dimension;
- encourage new sources of finance from the public and private sectors, especially by way of public equity and specialised private infrastructure funds;
- provide technical assistance to project promoters;
- foster the development of transnational operating companies;
- encourage introduction of user charges on a consistent basis across the EU;
- provide a framework for the exchange of information and the settlement of conflicts;
- coordinate the pooling of non-commercial risk and broker appropriate insurance cover;
- render independent advice to the EU institutions.

We recommend that this proposal is included on the agenda of the current Intergovernmental Conference which is already examining the parallel questions of an EU competition agency and the associated question of EU company law.

12. In the longer term, there is a strong case for endowing the Agency with limited executive powers. It would then be able to arbitrate disputes within partnerships, and, in conjunction with the European Investment Fund, encourage the replacement of grant finance with public equity thereby creating the basis for more private participation and the development of a European Infrastructure Fund.

13. At the member state level, especially in France and the United Kingdom, public private partnerships are demonstrating how they can pioneer the provision of infrastructure. In the nineteenth century the private sector was the leading provider. In the twentieth century the public sector took over the role. In the twenty-first century, guided by a European Union Infrastructure Agency, public-private partnerships could become the central provider of infrastructure at the European level.

1 | The Challenge

One of the most vital foundations of Europe's economic prosperity is modern infrastructure. Much of what we have is obsolescent in terms of technology and fractured by state boundaries and, accordingly, unfit for the 21st century. To make Europe more dynamic and competitive in world markets, the European Union embarked in the 1980s upon the great project of building a single market in which freedom of movement of goods, money, services and people would be assured. In the 1990s the concept of the Trans-European Networks (TENs) has emerged as a complement to that of the single market.

Much of the inspiration behind TENs has come from industry. Europe's leading industrialists complained that the state of the infrastructure did not permit full advantage to be taken of the single market. In 1984, the European Round Table of Industrialists (ERT), in its acclaimed report *Missing Links*, appealed to governments to think of the European economy as a whole, as industry did in its daily business, and to match Europe's infrastructure to the new reality. In particular, the ERT identified serious gaps in cross-border ground transport that were the legacy of narrow, national planning. When coupled with problems of geography, the report argued, most of these problems required costly solutions that were bold in terms of technology.[1]

This Report argues that the European Union (EU) needs to invest heavily in its public utilities of transport, water and energy supply, and information technology infrastructure, and that this investment requires new models of partnership between private enterprise and the public sector. These partnerships will encompass several different sources and types of financing, by far the greater part of which will be generated from a variety of private sources. Restrictions on public finance are already self-evident, and are being further tightened by the imperative for member state governments to meet and then sustain the EMU convergence criteria set by the Treaty of Maastricht. Likewise, growth in the EU budget is constrained by the 'financial perspective' until 1999 of 1.27% of EU GDP.

Changing industrial environment

At the same time, the steady progress of liberalisation, increased competition especially in public procurement, and restrictions on state aids are combining to transform the public utility sector. This process has advanced farthest in the UK, where privatisation has already affected the telecommunications, railways, water, gas, electricity and nuclear industries.

The traditional approach of the public sector has been to seek value for money by separating the design, construction and operation processes in successive lengthy and bureaucratic tender procedures. This has tended to discourage the private sector from making an effective contribution, especially in terms of timescale and cost-effective design. The public works mentality that predominates in many European countries has discouraged entrepreneurial risk capital and has led the private sector to seek substantial, even perhaps excessive guarantees for public sector involvement, which, in turn, have defeated the purpose of the public-private partnership.

Public enterprise

It is clear, therefore, that while the provision of European infrastructure cannot simply be left to market forces it does require a reform of bureaucratic attitudes within public authorities at all levels. The private sector cannot perform without enterprising political decisions on routing, technical standardisation, universal access, environmental assessment and strategic planning. Most of the TENs transport projects, for example, pursue broad public objectives as well as those of European integration and sustainable growth. Similarly, the delivery of broadband communications to the more peripheral parts of the Union is vital if the network is to bring widespread educational and other social benefits. In both cases, public funds will therefore be required to broker the marriage between political aspirations and economic viability.

The building of TENs also required a common approach from governments within the European Union. A common transport policy had been one of the elements in the original Treaty of Rome, but it has had only modest results. At any rate a broader approach, reflecting the changed industrial and political environment, was badly needed. The Treaty on European Union, signed at Maastricht in 1992, provided a fresh start.

What the Treaty says

The Treaty of Maastricht formally recognised the importance of the missing links by including a new section, Title XII, on Trans-European Networks. Article 129b says that, in order to derive full benefit from the single market, 'the Community shall contribute to the establishment and development of trans-European networks in the areas of transport, telecommunications and energy infrastructures'. It continues:

> 'Within the framework of a system of open and competitive markets, action by the Community shall aim at promoting the interconnection and interoperability of national networks as well as access to such networks.'

Peripheral regions should receive special attention. In order to achieve these objectives, Article 129c entrusts three main tasks to the Community:

1. to 'establish a series of guidelines covering the objectives, priorities and broad lines of measures envisaged in the sphere of trans-European networks; these guidelines shall identify projects of common interest';
2. to 'implement any measures that may prove necessary to ensure the interoperability of the networks, in particular in the field of technical standardisation';
3. to 'support the financial efforts made by the Member States for projects of common interest financed by Member States ... particularly through feasibility studies, loan guarantees or interest-rate subsidies'.

7

The Community may also contribute to transport projects in the four poorer member states of Spain, Portugal, Ireland and Greece via the Cohesion Fund. The Treaty adds pointedly that the Community's own activities 'shall take into account the potential economic viability of the projects'. The framework guidelines shall be adopted by a co-decision of the European Parliament and the Council, acting by qualified majority vote (QMV). [2]

Member States shall coordinate their national policies, and the Commission may 'take any useful initiative to promote such coordination'. Cooperation with third countries is foreseen. These coordination and international activities shall be decided by QMV in the Council, unless the Parliament has opposed the measure and is supported by the Commission, in which case unanimity is required. [3] In any case, any member state may veto unilaterally any project that affects its own territory.

The Delors White Paper

As soon as the Maastricht Treaty had come into force, the European Commission wasted no time in giving TENs a high priority. The Commission White Paper *Growth, Competitiveness, Employment* was published in December 1993. [4] An important facilitator of the single market was to be the accelerated establishment of trans-European infrastructure networks.

First, the White Paper called for the creation of an information society. The USA and Japan had taken the lead in the new technological revolution. Europe's main handicaps were the fragmentation of telecoms markets and the lack of major interoperable high-speed, large capacity links. Here, the Commission argued, European-level partnerships between the public and private sectors would succeed in creating a networked information society in Europe. The objectives were:

- building a system of infrastructures (cable and land or satellite-based radio communications, including integrated digital networks);
- developing services (electronic images, databases, electronic mail);
- promoting applications (teleworking, teletraining, telemedicine and linked administrations).

8

The second theme of the White Paper was the promotion of new or better-designed trans-European transport and energy networks. The Commission described how Europe's investment in infrastructure had slowed over the last decade. This was particularly true of transport, with resulting rigidities, procedural slowness and malfunctions. The White Paper and subsequent studies, such as the work of the Ciampi Committee on competitiveness, have suggested that poor transport is one of the principal reasons for Europe's sluggish economic performance. [5]

The White Paper suggested that the level of investment required before the end of the century was Ecu 250 bn. The Commission prioritised certain transport TENs:

- new strategic trans-frontier links, including the Brenner rail link, the Lyons - Turin rail link, the Berlin - Warsaw - Moscow motorway link;
- improved interconnections between various transport modes, for example, the Heathrow - London - Channel Tunnel link;
- improved interoperability and efficiency of networks by traffic management systems (air, sea, land) which would significantly reduce nuisance factors.

The White Paper recognised that the transport sector both required the most substantial investment and suffered the largest gap between need and availability of resources. But the Commission was determined to establish the best possible conditions for the financing of transport by way of partnerships between the public and private sectors.

Financing the networks

The White Paper argued that the TENs in telecoms and energy could be financed mainly by private investors. It was estimated that Ecu 67 bn would be required for telecoms networks between 1994-99 of which the Community could provide Ecu 5 bn from the R&D framework programme and from the structural funds.

Over the same period, the Commission identified a series of 25 transport priority projects amounting to Ecu 83 bn and ten energy projects costing Ecu 13 bn covering the Union but also extending to Central Europe and North Africa. It further identified environmental programmes of sufficient size to merit eligibility for financial support from the Community: these concerned urban waste water treatment and renovation of water supply distribution systems at an estimated cost of Ecu 140 bn by the end of the century, of which the European Union (EU) could help to finance Ecu 25 bn over the period 1994-99. Estimates on an annual basis are shown in Table One.

TABLE ONE

Annual Financing of TENs 1994-99

Ecu bn

European Community budget		**5.3**
TENs	0.50	
Structural Funds:		
TENs	1.35	
Environment	0.60	
Cohesion Funds:		
TENs	1.15	
Environment	1.15	
R&D:		
Telecoms	0.50	
Transport	0.05	
EIB loans		**6.7**
Union Bonds (mainly transport & energy)		**7.0**
European Investment Fund (mainly telecoms)		**1.0**
TOTAL		**20.0**

Source: Growth, Competitiveness, Employment

The White Paper proposed that the major portion of finance, about Ecu 420 bn, should be raised at the level of the member state either through private investment (especially in the telecoms sector) or via public enterprises. The European Union would play a role, as foreseen in the Maastricht Treaty, by supporting the financial efforts of member states for TENs through feasibility studies, loan guarantees or interest rate subsidies. Finance available directly through the Community budget would total Ecu 26.5 bn over 1994-99, leaving Ecu 73.5 bn to come from the European Investment Bank and new financial instruments, such as the proposed 'Union Bonds' and the European Investment Fund.

Union Bonds

The Delors Commission advocated the utilisation of Union Bonds as a mean of raising long-term, fixed interest loans on the international capital markets. The beneficiaries would be project promoters directly involved in TENs, including public sector agencies and private companies. The European Investment Bank (EIB) would be invited to appraise and advise the Commission on the overall structure of the bonds, and to act as agent for individual loan contracts.

In the event, the European Council in Brussels in December 1993, which otherwise accepted the White Paper, rejected the proposal for Union Bonds on the grounds that the EIB was already able to borrow on the best terms and had more expertise than the Commission of lending for infrastructure projects. Instead, the Council set in hand three parallel exercises with respect to the European networks:

- the establishment of a group of personal representatives of the leaders under the chairmanship of Commission Vice-President Henning Christophersen to guide and accelerate the work on transport and energy networks;
- the creation of an *ad hoc* group of high-level experts under the chairmanship of Commissioner Bangemann to report on the information society;
- a study of the obstacles to the financing of TENs in all these sectors and of the major environmental projects. The object of the study was to see how the EIB could raise additional loans of Ecu 8 bn a year to meet the needs of the TENs without undermining efforts to reduce public debt. Emphasis was also laid on the need to mobilise larger amounts of private finance for these projects by reducing their financial risk.

Christophersen group

The Christophersen group reported to the European Council at Essen in December 1994, having given a preliminary report at Corfu in June. It emphasised the need to prioritise the TENs and to ensure that they passed the test of economic viability. Priority TENs should be expected to produce a substantial net benefit taking account of the external as well as the direct costs and benefits. They should contribute positively to the competitiveness and the technological development of the EU economy. In the transport field, it was recognised that not all projects would be viable in strict financial or commercial terms, because revenues would be insufficient both to cover costs and to produce an adequate return to investors. The estimated financial rates of return for transport projects ranged from 3 - 8%, which implied that some form of public support would be required unless the external costs and benefits could be captured by way of user charges or other revenue generating mechanisms. [6]

The Christophersen group proposed various transport TENs (see Table Two and Map One) which, it claimed, enjoyed the following characteristics:

- common interest, such as cross-border sections;
- large scale, bearing in mind the type of project and the relative size of the member states concerned;
- economic viability and scope for private involvement;
- contribution to Union objectives such as economic and social cohesion; respect of other Union policies, notably on environmental protection;
- maturity.

Despite their best intentions, it is far from clear that the Christophersen group achieved a consistent analysis. Different member states have very different approaches to transport projects, especially with regard to high speed rail. They have different criteria for choosing projects in the first place, as well as to their subsequent evaluation; they differ about the calculation of 'public good' in terms of socio-economic values, and in their assessment of short and long term benefit (see Table Nine, page 87).

Christophersen, indeed, recognised that there were formidable problems in the path of implementation of the priority projects. Each project had its own characteristics; what they had in common was that they shared one or more of the following features:

- cross-border nature;
- large scale;
- moderate financial viability;
- high risk.

TABLE TWO

Priority TENs Projects

Transport Projects

1. High Speed Train/Combined Transport North South (Verona-Brenner-Berlin)
2. High Speed Train (Paris-Brussels-Cologne-Amsterdam-London)
3. High Speed Train South (Madrid-Barcelona-Dax/Montpellier)
4. High Speed Train East (Paris-Strasbourg-Karlsruhe)
5. Conventional Rail/Combined Transport (Rotterdam-Ruhr)
6. High Speed Train/Combined Transport (Lyon-Turin-Milan-Trieste)
7. Greek Motorways: Pathe and Via Egnatia
8. Motorway Lisbon-Valladolid
9. Conventional Rail link Cork-Dublin-Belfast-Larne-Stranraer *
10. Malpensa Airport (Milan) *
11. Fixed Rail/Road link between Denmark and Sweden (Øresund fixed link) *
12. Nordic Triangle (rail/road)
13. Ireland/United Kingdom/Benelux Road Link
14. West Coast Main Line (rail)

* Projects with secured finance according to Christophersen group criteria.

Energy Projects

1. Greece-Italy electrical interconnection (submarine cable)
2. France-Italy electrical interconnection
3. France-Spain electrical interconnection
4. Spain-Portugal electrical interconnection
5. Denmark East-West electrical interconnection (submarine cable)
6. Greece natural gas network
7. Portugal natural gas network
8. Portugal-Spain gas interconnection
9. Algeria-Morocco-European Union gas delivery pipeline
10. Russia-Belarus-Poland-European Union gas delivery pipeline

Source: Christophersen Report

Obstacles

There are five types of obstacle that the EU's priority TENs projects have to overcome:

1. *Political obstacles*, such as the difficulties in reaching clear and lasting agreement among all the public authorities involved about the timing and evaluation of the external effects, necessary compensatory mechanisms, technical features, construction and operation of the networks;

2. *Regulatory framework obstacles*, relating to administrative, regulatory and legal procedures that provide the operating environment for project planning, construction and operation;

13

3. *Financial obstacles*, relating to drawing together the business plan for the entire project in all its phases, the timing and conditions of financial commitments, and the insufficient availability of public funds;

4. *Cultural obstacles*, namely the long traditions of the public financing of transport infrastructure, combined with reluctance to tackle public-private partnership schemes;

5. *Methodology obstacles*, in that there is no agreed common approach on how to recognise or evaluate the European dimension.

The cultural problem should not be overlooked. Transport authorities are usually ill-suited to attracting private partners. Although it may be generally admitted that the private sector could introduce better management techniques and technological know-how into the provision of major infrastructure projects, the entrepreneurial element is often deterred by the officious administrative and regulatory procedures of the public sector. Those railways which have newly acquired corporate legal status may lay claim to being entrepreneurial, particularly if they have separated their infrastructure from their operations. But unless they have been privatised the burden of servicing their large debt is still a call on public expenditure.

With regard to regulatory obstacles, Christophersen recognised that there was a need for 'horizontal' measures that could be taken between the member state governments to streamline national approval procedures and to improve technical interoperability, 'but also in providing for increased flexibility in the form of public financial support extended to TEN projects for the purpose of facilitating private involvement'. [7] The total sums involved remained in the estimation of the Christophersen group very large, with about half the total investment of Ecu 91 bn required by 1999.

Where user charges can be utilised it will clearly be easier to build a viable partnership with the private sector. But long, uncertain and expensive construction periods must be financed before revenue comes on-stream. Involvement of the private sector at an early stage is therefore also needed in order to improve cost-effectiveness in project planning and construction. There will often remain a further financing gap between the *economic* return, assessed by cost-benefit analysis, and the *financial* return. In the past the state would have

MAP ONE: PRIORITY TENS

KEY

——— Rail

– – – Road

·········· Both rail & road

–··–··– Nordic Triangle

filled this gap on a large scale through grant-aid or guaranteed bonds. But we have already noted the current constraints on public expenditure in every member state of the Union; and the discipline of the single currency regime will continue to prevail. **Public-private partnerships in the transport sector are not a luxurious optional extra but an essential requirement.**

The Essen European Council

The European Council at Essen in December 1994 accepted the Christophersen report and approved the proposed priority projects (Table Two). It emphasised that the obstacles were mainly legal and administrative. It called upon the Council of economic and finance ministers (Ecofin) to 'top up the funds currently available' for the TENs. [8] The leaders also encouraged the European Investment Bank (EIB) to enhance its support for TENs beyond that agreed at the Edinburgh European Council two years earlier. A TENs 'window' was created for lending to the public and private sectors and to partnerships between them. We return to the role of the EIB in Chapter Four.

INFORMATION NETWORKS

The high level group of industrialists chaired by Martin Bangemann to consider the measures needed to install information infrastructure throughout the Union reported in May 1994. It urged the EU to 'put its faith in market mechanisms as the motive power to carry us into the Information Age'. It went on:

> 'This means that actions must be taken at the European level and by the Member States to strike down entrenched positions which put Europe at a competitive disadvantage:
> - it means fostering an entrepreneurial mentality to enable the emergence of new dynamic sectors of the economy;
> - it means developing a common regulatory approach to bring forth a competitive, Europe-wide, market for information services;
> - it does NOT mean more public money, financial assistance, subsidies, *dirigisme*, or protectionism'. [9]

The Commission emphasised the crucial importance of liberalising the telecoms market so that infrastructure projects would flow from social and market demand.

16

In 1995 the Federal Trust published a report entitled *Network Europe and the Information Society* based on the work of a study group chaired by Lord Cockfield. [10] It looked particularly at three aspects:

- the creation of the infrastructure;
- the development of the services using that infrastructure;
- giving widest possible access to those services.

It is only if all three are addressed that the full benefits of the Information Society can be harvested. And they need to be addressed not just in the context of the existing European Union but of an enlarged one.

The *Network Europe* report found that the research and development of the science and technology involved in designing 'information superhighways' are not national; neither is the investment required to build them; nor are the markets to exploit them. Indeed, it is in the networking of the *Infobahnen* across the continent and globally that much of their value will be realised. The issues facing European governments are how to promote the investment required to build the Europe-wide infrastructure within a short period; and how to mobilise the new technologies to enhance competitiveness, the quality of public services, access by the widest possible spectrum of the public and in general to contribute to improving the quality of life for the people as a whole. Until now, the European Union has been mainly concerned with issues of competition and infrastructure. These will continue to be important, but the EU must be more dynamic in using the emerging Network Europe to promote education and information.

The potential demand for services is very great — but public policy errors could delay their development by years, or put them beyond the reach of most citizens. The European Union has established an important role, both regionally and internationally, in the development of a networked Information Society. The EU needs to accelerate its programme to open existing telecommunications monopolies and infrastructure to effective competition, and to insist on its completion. Member states must establish independent regulatory structures well in advance of the EU's liberalisation deadline of 1 January 1998. Until policy over competition and regulation is clarified, the questions of financing, licensing and interconnection will remain problematical.

Regulation

The Federal Trust study group did not think that the case for a single EU regulator in the IT/telecoms sector had yet been made out. The study group proposed, instead, that a formal committee of national regulators be set up, chaired and serviced by the European Commission, to promote best practice between member state governments and to tender advice to EU and national competition and regulatory authorities.

If the Information Society is to become a reality soon, the European Union must resolve its approach to competition and universal service with regard to the new broadband network.

> *Competition.* As liberalisation advances, the general principles of EC competition law on abuses of a dominant position and restrictive agreements, the rules on the free movement of goods and services, and the restrictions on state aid are, if strictly applied, adequate for the development of broadband networks and new services using them.

> *Universal Service.* It is vital that the desire to guarantee universal service does not unduly hinder investment or distort competition. The Cockfield group recommended the definition not of minimum standards of service which must be provided in all areas, but of maximum levels of service on which member states can insist. To promote access and services in uneconomic areas, it proposed the establishment of national universal service funds, managed by national regulators and supported by a strict approach to the separation of costs in the accounts of operators and to the application of EC Treaty rules.

Widespread access is the key. The immediate priority is to ensure that services are widely available at an early stage, and that attention is paid to the needs of both the major corporate users and the individual citizen. The imposition of formal 'universal service obligations' should be progressive, taking account of the cost and availability of new services.

The principles of Open Network Provision (ONP) need to be emphasised. EC law already sets out a series of basic requirements on access to the network. Conditions of access are to be transparent and, above all, equal: the operator cannot privilege his own operations over those of other service suppliers. This applies not only in such matters as access charges but also in the software access protocols. The ramifications of ONP for commercial operations are evident. The principle of providing access to the superhighway will be a basic building block of the future European economy.

Public funding at the EU and national levels should be tied to the development and take-up of services on the network, particularly among those users, notably in the health and education sectors, that are largely dependent on the state. Such funding should be directed to the users not the suppliers. The EU could have a dynamic role to play in promoting pan-European educational and information services along the superhighways. The EU structural funds should be redirected towards the same goals, especially in the peripheral regions. The EU's instruments may need to be adapted further to meet the financing needs of the Information Society as new stages in its development are reached.

Investor confidence in Network Europe will grow once the political conditions are set, the regulatory framework is in place, and the pace of technical standardisation increases. To encourage investment and innovation, bans on telecoms operators providing certain forms of service, such as broadcast entertainment, should be only temporary.

The Cockfield group believed that the successful combination of building the information network and promoting its use throughout society will have six highly beneficial consequences:

- the task of building the EU Single Market will be completed by the addition of the physical infrastructure;
- business will use the capability of integrated information technology and telecommunications to become more productive and competitive;
- government at all levels will exploit the same facilities for public administration;
- many jobs — perhaps ten million — will be created once Europe's IT manufacturing and service industry gears up to satisfy consumer demand;
- both the liberalisation and integration of Central and Eastern Europe will be hastened by the information technology revolution;

- European citizens will be empowered by greater access to education and information to take more charge over their own lives.

It is difficult to escape the conclusion that progress towards the building of TENs in IT/telecoms is faster and more determined than in the transport sector because of the much higher level of private sector involvement. As the Commission reported to the Madrid European Council in December 1995, the 'bottleneck in the field of telecommunications concerns mainly the interoperability of generic services and the development of applications'. [11] A number of challenges remain to be met:

- raising awareness of users to the advantage of the Information Society;
- building the public-private partnerships to realise services and applications in response to public needs;
- limited budgetary resources requiring maximisation of their leverage effect.

ENERGY NETWORKS

Trans-European energy supply networks are rightly regarded as an important contribution to enhancing competition. The Christophersen group took the view that obstacles to the development of energy networks were generally less than in transport. However several priority energy projects are still not completed. The Italy/Greece, France/Italy and France/Spain electricity interconnections are all having problems relating to planning authorisation. Problems of economic appraisal and/or financing are still affecting, amongst others, the Italy/Greece and the Denmark East-West electricity connection and the Greek natural gas project.

The Council and the European Parliament have agreed guidelines for energy TENs. They seek to identify projects which will connect isolated regions to natural gas and electricity networks both within the periphery of the EU and in Central Europe and the Mediterranean. They are aimed at encouraging the rational production, distribution and utilisation of energy resources. [12]

For these and other potential energy projects the regulatory environment is the key element in establishing the appropriate incentive for investment. The Electricity Directive, agreed after much debate in 1996, marks a modest step in the direction of a single market, although only the largest consumers are affected, leaving the bulk of electricity demand in the hands of monopoly local suppliers.

A Gas Directive is now under discussion and seems likely to follow a similar course. Although both Directives do represent forward steps, greater progress in establishing a single market in energy will be necessary if the full potential of energy TENs is to be realised. In the meantime, consumers in many European countries will continue to pay higher fuel prices than necessary with consequent damage to Europe's competitiveness.

Slow progress

In the energy as well as the transport fields, therefore, progress in building the TENs has been disappointing. The Commission complained frankly to the Madrid European Council a year after Essen that:

> 'Although the examination of individual priority projects shows substantial scope for enhancing the involvement of the private sector, very few public-private partnerships are being set up.'

Member state governments have themselves delayed funding for at least two of the fourteen priority transport projects, creating a shortfall of Ecu 760 m. This may suggest that the decision of the leaders at Essen to select the biggest and most prestigious projects was precipitate and over-ambitious. (Christophersen had also offered a list of more modest proportions which may have represented a better starting point: see Table Three.)

The challenge

The Director of the European Centre for Infrastructure Studies, Wolfgang Hager, has pointed out that with many of the key transport TEN projects the European Union is entering three uncharted territories at once: public-private partnerships, genuine cross-border projects; and technological complexity so far only matched by the Channel Tunnel railway.

21

'However, the rewards from doing so are substantial. Indeed, without the extra speed and cheaper delivery, some of the most interesting projects may suffer the fate of so many infrastructure projects: we shall still be discussing them in twenty years time. If on the other hand, we dare to re-invent an important part of the mixed economy, managing greater productivity with a collective, imaginative commitment in the future, Europe will have gained a symbolic as well as a material achievement of which we can be proud'. [13]

In subsequent chapters we look at ways to make such a 'mixed economy' work; and we make recommendations, particularly in the field of transport.

TABLE THREE

Further Projects of Importance

Acceleration possible so that work can begin in about two years

Combined transport projects identified up to now in France, Germany, Italy, Belgium, Portugal and Spain
Sparta airport (Athens)
Berlin airport
Maurienne motorway
Marateca-Elvas motorway
High-speed train in Denmark
Transapennine highway Bologna-Florence
High-speed train/combined transport Danube axis
Munich/Nuremberg-Vienna-Budapest-Bratislava
Nice-Cuneo motorway

Projects which need to be further examined

Fehmarn Belt fixed link
Bari-Brindisi-Otranto motorway
Rhine-Rhône link
Seine-Scheldt link
Elbe-Oder link
Danube upgrading between Straubing and Vilshofen
High-speed train Randstad-Rhine/Ruhr
Amsterdam-Arnhem-Cologne
Road corridor Valencia-Saragossa-Somport
High speed train Brenner-Milan-Rome-Naples
Magnetic levitation train: Transrapid (Hamburg-Berlin)
High-speed train connection Luxembourg-Brussels
Road corridor Naples-Reggio di Calabria

Source: Christophersen Report

---

2 | Partnership in Practice

In this chapter we look at some different national experiences of public-private partnerships.

FRANCE

France's experience of state involvement with private entrepreneurs in the fields of water, lighting and transport stretches back to Colbert in the seventeenth century when canals and bridges were built by granting concessions.

The advantage of the French model has been that comprehensive integration can be achieved in the design, construction, operation and maintenance of a facility or service. The concession can be granted before all the details are known. Tenders are decided on the basis of the concept and the track-record of the bidders. This approach is facilitated by a system whereby key officials in both the public and private sectors are likely to be graduates of the *grandes écoles*.

The French approach to public subsidies is also very pragmatic. The public authority relies on independent analysis to propose an economically efficient tariff; and then to calculate the revenue needs of the private sector. Any gap between the two is then filled by subsidy provided that this is not greater than the additionally computed socio-economic benefit. The French place great importance on the long-term nature of partnerships and consider that the fundamental notion is one of trust between the partners. According to Claude Martinand:

> 'The best analogy that can be made is that of a marriage contract which does not depend on the number of potential

suitors or on the initial dowry. Instead, the success of this contract depends on the harmony which prevails between the partners for as long as they are together and which ensures their well-being'. [14]

This long-term approach is exemplified by the duration of the concessionary period: the Normandy bridge concession extends to 50 years, highways from 30 to 45 years. A commentator has suggested that the French and British systems have been developed from radically differing attitudes to the duration of contracts and on the regulatory methods employed. Dominique Lorrain writes: 'They differ in all respects, even including their expectation from a cultural standpoint, the level of trust or mistrust in the public-private relationship. To some extent, the British system depends on distancing and suspicion. The worlds inhabited by local authorities and private business never meet'. [15]

Cofiroute

The French government introduced toll operation of highway projects in 1955. However, the five concessionaires created following this legislation were public-private ventures with low levels of capitalisation and under the ownership of local authorities whose role was limited to collecting tolls. Until 1970 the state continued to act as the primary contractor for the construction and maintenance of the highways, at which point new legislation opened up highway concessions to private investment. It was recognised that only a relatively long concessionary period of 35 years would provide the necessary level of profitability. As the concessionaires bear the responsibility of traffic forecasting, it was felt necessary that they should be able to set the tolls independently within a contractual ceiling indexed to inflation. In fact, the French government has never complied with this arrangement and has set the level of tolls year after year under its separate price-controlling powers.

Four private companies were created at the time of the 1970 reforms but only Cofiroute survives. The other three were severely hit by the effects of the first oil shock and after ten years had to appeal for government backing with the result that they were re-nationalised. The state provides a guarantee to Cofiroute that should the concession run into financial difficulties it will meet in full any payments due from the concessionaire to the lender.

The French authorities have adopted a flexible approach to the sharing of any surplus profits carried over and above the expected rate of return. The surplus provides partial self-finance for subsequent projects. This financial transfer mechanism underpins the French development of roads, water supply and public transport. Otherwise, as Jean-François Poupinel has pointed out, there is a temptation for elected officials overseeing the concession to forget that the profit and loss account only turns positive after a period of 10 to 15 years. [16] Risks are greatest near the beginning of a project as a result of cost overruns during the construction period and insufficient traffic levels at the start-up of operations. Expensive repairs will also affect traffic levels: Cofiroute has a strong interest in ensuring that very high standards of road-building are achieved — which may help to explain why French motorways do not seem to be as plagued by cones as their British equivalent.

THE UNITED KINGDOM

In the United Kingdom there has been a radical change in attitude over the last decade to the use of private finance to fund investments by public sector bodies. Until 1989 policy was governed by a Treasury orthodoxy encapsulated in the so-called 'Ryrie Rules' of 1981. By 1988 they had been refined to two 'fundamental principles':

- privately funded solutions must be tested against publicly funded alternatives and shown to be more cost-effective;
- unless ministers decided otherwise in particular cases, privately funded projects should not be additional to public expenditure provision — in other words, provision for public expenditure would be reduced by the amount of private funding obtained. [17]

During the 1980s, the Ryrie rules acquired the status of Treasury bogy. They certainly held back the government's programme of Urban Development Grants. David Willetts MP, who as a junior Treasury official helped to formulate and enforce the rules, has written:

'The Treasury's objective then, though not always openly stated was to stop such schemes [public-private finance]. The notorious Ryrie rules were a tease — the conditions

25

they set for private financed projects were not intended to be met in practice'. [18]

A significant relaxation was introduced in 1989 when the 'additionality' restriction was substantially eased.

Urban Development Grants

Urban Development Grants (UDG) were introduced in 1982 and closely followed the model of the American Urban Development Action Grants (see below). UDG was intended as a limited capital subsidy that would induce or leverage additional net private investment. Following the US pattern, central government appointed a UDG appraisal team whose task was to evaluate, on the basis of the private sector risk investment proposed, the viability of the scheme, the practical feasibility of the project and the contribution the development would make towards improving economic, social

---

TABLE FOUR

The Dartford - Thurrock River Thames Crossing

1. This was the first major project to be completed under the auspices of PFI. It is financially free-standing, the only public sector contributions being the cost of the approach roads and the transfer of the revenue stream from the existing tunnel, which the supplier (the concessionaire) took over. The supplier designed, constructed and operates the bridge, and recovers the revenue from tolls. The concession period is for a maximum of 20 years or until the company has accumulative revenue which exceeds the amount of debt outstanding, whichever is sooner.

2. Although risk transfer to the private sector has been significant, it is subject to two important qualifications:

- as it is expected that the agreed cumulative revenue will be achieved well within the maximum 20 year concession period, in practice most of the demand risk is borne by bridge users and will be reflected in a shorter or longer tolling period;

- the bridge transfers to public sector ownership after the end of the concession period: if latent defects emerge after transfer, the state will be responsible for rectifying them.

Source: *Private Opportunity, Public Benefit*, H M Treasury

---

or environmental conditions. The British UDGs had only a limited success with a cumulative total of £44 m allocated by 1985/86.

Out of 514 project applications submitted to the Department of the Environment, only 22% were implemented. There were criticisms by developers of the long delays in getting approval from Whitehall (in Washington there was a guarantee that a decision would be made in three months). An independent UDG evaluation report suggested that the gap between concept and delivery was due to a combination of the haste with which the programme had been conceived, the inadequate skills of local government and the 'ineptness of the private sector in putting together cogent development propositions'. [19] A government report summed up as follows:

> 'There was a widespread consensus that the attempt to encourage the formation of partnerships has been a well-adjudged priority. Most of our discussion, however, suggested that it has been a goal which has been only imperfectly achieved in practice' [20]

The Private Finance Initiative

A new policy — the Private Finance Initiative (PFI) — was introduced in the autumn of 1992 when Norman Lamont, then Chancellor of the Exchequer, laid down that:

- self-financing projects undertaken by the private sector would no longer need to be compared with theoretical public sector alternatives;
- the government would actively encourage the private sector to take the lead in joint ventures with the public sector;
- the public sector would have greater opportunity to use operating leases.

There are two fundamental requirements for a PFI project. First, value for money must be demonstrated for any expenditure by the public sector. Second, the private sector must assume genuine risk.

The ideas underlying the PFI had their origins in projects with their own third party revenue stream — for example, estuarial crossings, such as Dartford, roads with real tolls, and other transport infrastructure. Such projects are the most directly analogous to public-private partnerships in European infrastructure projects. However, PFI has now gone a long way beyond its origins to the point where the public sector itself is to focus on projects and services in which the private sector's higher cost of capital can be outweighed by the benefits of:

- innovation, efficiency incentives and risk management skills as applied by the private sector to project design and operation;
- optimisation of whole lifecycle costs (free from the upfront capital constraints which inhibit the public sector); and
- the exploitation of any spin-off commercial opportunities.

PFI has to deliver a better value service, defined in terms of price, quality and risk reduction.

Value for money

The Treasury guidance on PFI places a high value on the competitive bid procedure:

> 'A critical question in deciding whether to go ahead with a PFI option is identifying the best value for money. Competition is the best guarantor of value for money. As a result of the competitive process, the best PFI options should emerge. These may involve comparison with a conventionally procured alternative — the public sector comparator'. [21]

However, some of the most strident criticisms of PFI heard by the House of Commons Treasury Committee relate to the competitive bidding process. [22] It appears that bids are often requested for projects that are either unsuitable for PFI, by virtue of their size and nature, or for projects that are unlikely to proceed. Government policy was to encourage departments to set up bids almost for their own sake. In November 1994, the Chancellor spread the practice originally adopted in the National Health Service whereby the Treasury would not approve any capital projects unless private finance options had been explored. [23]

The government claims that PFI has produced substantial savings, although it must be difficult to evaluate precisely because of the long-term nature of the projects. The Treasury asserts that, with regard to the supply of Northern Line tube trains, 'the train service contract is in excess of 20% better value for the public sector compared with the purchase alternative over 20 years'. [24] Evidence from Dr Stephen Glaister to the Commons Treasury Committee helped to elucidate the types of areas where savings are most likely to be derived:

> 'In the case of buses, part of the cost savings came about because the private sector used its labour more flexibly than had the public sector, while in construction, it is likely that the same firm will be building the assets, so the benefits will derive from design requiring low maintenance'. [25]

The Treasury Committee might have quoted another extract from Dr Glaister's evidence:

> '*Mr Nigel Forman MP*: But is not the key point there really the lifetime cost of the project?

> '*Dr Glaister*: That is right. It is looking at things as perhaps an aircraft manufacturer or the leaser of an aircraft would look at it, recognising that ten years or 20 years down the road one is going to have to do major maintenance work on this asset, it is going to have to be maintained. So you design it from the beginning in such a way that it can be done cheaply instead of forgetting about all of that and building something in such a way that it actually failed disastrously after five years and then had to have very expensive repairs. We see that all around us. We have seen motorways breaking up and having to be repaired after a very short period of time because not enough care was given to the long term maintenance'. [26]

Unbundling of risk

The theory behind the Private Finance Initiative is that **risk should be apportioned between the public and private sector according to where it can best be managed.** It is important to note that **obsession with risk transfer for its own sake is likely to undermine 'value for money' objectives.** This point has been understood only recently and widely within the public sector. However, concerns about whether a project will be classified as public sector or private sector (with accounting conventions looming large in this context) still provide the potential to trip up projects.

On the other hand the 'unbundling' of risk has revealed that there are important hidden costs traditionally borne by the public sector. An example is provided by the public-private prison deal, where the two main risks to be transferred to the private supplier are:

> *Design and construction risk.* Payments begin when cells are ready for occupation. If there is a failure to provide the required number of places, the supplier is liable to pay liquidated damages.

> *Availability and operating risk.* Once the prison is operational payment is only made for places that are available for use. Payment may also be reduced if the supplier's performance falls below specified levels (if for example there is an escape). The supplier also bears the risk of places being unavailable as a result of gaol disorder.

The good news is that the 'Design, Construct, Manage and Finance' (DCMF) prisons have been built to such a superior design to traditional state prisons that escapes and riots have been virtually non-existent. [27]

Commitment and accountability

The Treasury Committee recognises that flexible project specifications can increase the element of innovation and that there is accordingly a need to protect intellectual property rights. The Committee is concerned that open-ended specifications may make quantitative comparisons more difficult and increase the cost of tendering. Their conclusion is:

> 'There is clearly a trade-off between greater flexibility of specification to encourage innovation and new approaches and the greater objectivity and precision that is possible in tender evaluation when specifications have been more standardised. We would not, on balance, wish procedures

to be adopted that vitiate the opportunity for new thinking which we see as one of the major attractions of PFI. We therefore feel that there may be a need for higher level scrutiny of such bidding processes, to ensure openness and protect against corruption'. [28]

For projects to have a chance of reaching successful conclusion two things must surely exist:

- strong political commitment to see the project through and sweep away the resistance of vested interests;
- the underlying financial case must be robust, and, if it is only robust with the addition of public sector subsidy, then the economic case for this subsidy must also be robust.

Projects which have strong political support but are not fundamentally viable and projects which have little political commitment despite having a good economic basis are both likely to waste a lot of time. The problem is that political sponsorship is often divided across different institutions or different parts of government. It is important to keep all of these interests moving in step: witness the effect on the Docklands Light Railway Extension of the Department of Transport's decision, half way through the procurement process, to reduce the cap on rail fare increases.

Off-balance-sheet finance

PFI is sometimes criticised for being merely a form of 'off-balance-sheet finance'. As Coopers & Lybrand pointed out in their evidence to the Treasury Committee, this underlines the need for proper management of PFI and for strict financial monitoring both for individual projects and in the round. A second criticism is that risk transfer will lead to increased costs to the public sector. This can be refuted by the argument that cost falls where risk is allocated to the party who can best manage it. The key is to find the optimum degree of risk transfer that can be done at the best price. In this context 'more' risk transfer is not necessarily 'better'. [29]

To date, the impact of the PFI on investment levels has been limited. According to the government's medium-term financial 'Red Book' there has only been around £1 bn of capital expenditure under PFI in the three years to 1995/96. This is less than 2% of total public sector capital spending and about 0.4% of total UK investment. There are planned increases of around £7 bn by 1999 which would represent 12% of total public sector capital spending. It would seem that the increase in PFI spending will be counter-balanced by a drop in other public capital spending commitments. This means that if the majority of projects do not go ahead as part of PFI they will not go ahead at all. Coopers & Lybrand sum up as follows:

> 'The depressed state of the UK construction industry at present and the evident need for investment in some areas of our national infrastructure (eg. railways, social housing) highlights the importance of removing the remaining barriers to the PFI and targeting new initiatives on areas where private finance will be additional to existing capital spending plans'. [30]

New skills needed in public sector

If PFI is to gain further momentum, it is essential that the civil service acquires the necessary deal-making skills. Unlike France, the British civil service has been compartmentalised from commerce and industry. **Decisions about whether a PFI solution is acceptable are based on two entrepreneurial tests: value for money and risk transfer. Both are difficult measures and have to be applied subjectively.** There are no set rules. Any decisions are open to later challenge by watchdog bodies — the Public Accounts Committee, the National Audit Office or the Audit Commission — all of which can use the benefit of hindsight to help determine whether the decision was the right one or not. As Coopers & Lybrand point out:

> 'This psychological barrier for our public servants, and the effort that will be required to overcome it should not be underestimated. In the short-term there may be little alternative to buying in these skills from outside the Civil Service. Further significant investment is required in PFI centres of excellence but this will only be effective if steps are taken to clarify responsibilities to produce an environment in which risk-taking is more acceptable in appropriate circumstances'. [31]

UNITED STATES OF AMERICA

The cultural gulf between the private sector and government, both federal, state and local, has never been as great in the USA as in the UK.

In America, public-private partnerships originated in grass-roots development in cities such as Baltimore and Pittsburgh. Civic and business leaders, appalled by the decline in business that accompanied inner city decay in the 1940s and 1950s, proposed that new development corporations should undertake comprehensive redevelopment. Baltimore and Pittsburgh had a strong tradition of civic culture and in both cases the business community took the lead in drawing up a blueprint or vision of the future. The mayor's office cooperated and powers were delegated to the respective development corporation allowing it to cut through the planning red-tape. The city authorities were responsible for land acquisition and site clearance. The private sector was responsible for finding developers and for identifying possible tenants for office buildings as well as raising private investment funds. Successful first phases led by the private sector were followed by more comprehensive redevelopment led by the mayor's office. The result is that both inner cities were transformed from drab, declining, industrial sprawl into bright, modern and dynamic cities that attracted new corporate headquarters and even considerable tourism.[32]

An authoritative study of such developments in several American cities defined public-private partnerships as follows:

> 'A public-private partnership is a sustained collaborative effort between government agencies and private organisations in which each of the partners share in the planning of projects and programmes designed to meet a public need and contribute a portion of the financial managerial and technical resources needed to implement those plans'. [33]

The UDAG initiative

The US federal government, recognising the importance of grassroots development, responded with a federal programme of Urban Development Action Grants (UDAG) in order to spread public-private partnerships nationwide. In recent years emphasis has switched from social fields, such as better housing, to economic development. State and city government are encouraged by Washington to be entrepreneurial and to take equity stakes in the developments.

The relative success of the UDAG programme in the 1980s can be attributed to the fact that it is market-related and flexible in purpose. Requests for UDAG support were submitted jointly by developers and city officials. UDAG staff examined the financing package and associated cash flow projections and often made suggestions about financial gearing and social targeting. An important factor in the selection of projects was the leveraging ratio, that is, the ratio of private to UDAG funds. By the end of 1983 a total of $17.2 bn private sector funds had been mobilised as a result of $3 bn of UDAG financing plus $1.6 bn from other government sources. [34]

A wide range of uses could be made of the Urban Development Action Grant:

- soft loans to a developer;
- land write-downs, demolition, relocation costs, on-site improvements and similar subsidies to reduce the developer's costs;
- infrastructure or other public improvements adjacent to a project site.

Rates and terms of the loans were related to the private sector's financial capability; loans were typically at below market rates of interest and were usually secured by a second mortgage on the project's property. There may well have been a bank loan or private bond issue as well. UDAG projects could include equity 'kickers' to ensure that some of the capital profits flow back to the city.

The UDAG programme was run down in the second half of the 1980s as the Reagan administration attempted to cut the overall federal deficit. Instead, Enterprise Zones were favoured, a concept borrowed from Britain, as a more ideologically acceptable means to generate new private investment. Advocates of Enterprise Zones tended to take the view that government intrusion into the economic life of inner cities was the root of urban distress. Enterprise Zones would promote a free market environment in designated urban areas by reducing taxes and relaxing or eliminating regulations on business activity. In the event, the federal initiative for Enterprise Zones failed to get enough support in Congress, although it was taken up by some states.

Empowerment Zones

The Clinton administration has moved away from the 'trickle down' theory. Nine 'Empowerment Zones' and 95 'Enterprise Communities' are to receive $1.4 bn in flexible block grants and $2.5 bn in tax incentives. To qualify localities have to meet certain poverty criteria and create strategies for revitalisation. The cornerstone of every application is that it must be based on public-private partnerships of community residents, state and local governments, businesses and financial institutions. The partnerships have to prove that local people are involved and that they can generate substantial additional investment. For example, in the successful Baltimore Empowerment Zone, $8 in outside resources was pledged for every $1 of federal funds. [35]

Importance of decentralisation

On both sides of the Atlantic, public-private partnerships are rightly seen as critical to urban regeneration. The American experience indicates that a decentralised approach produces better results. The business community have to be persuaded that they are full partners and that viable developments will not be frustrated through bureaucratic delays at central government level. It is interesting to note that the most successful public-private partnerships in the UK have been in Glasgow and Newcastle where there are strong regional or national identities.

In the US the role of the city and local authorities has always been very important, with a second stage of inner city projects being carried out under the leadership of the mayor's office. It is now generally accepted even in the UK that local authorities should be given more scope to play a role in the PFI. As the Department of Environment concluded in 1994:

> 'The impatience which central government showed with the unwillingness or inability of some local authorities to 'deliver' urban regeneration in the early years of the 1980s may well have justified the policy response of by-passing local government. But local authorities are generally now

seen as having an important role to play in the development and implementation of local strategies, especially now that an increasing number of such authorities have demonstrated their readiness to work in cooperation with a range of other partners'. [36]

Tax-exempt bonds

The US practice of combining tax incentives to stimulate private interest with municipal bonds to raise funds for the public expenditure makes the American process far less time-consuming and wasteful than the British. The federal government's largest contribution to the financing of public infrastructure (other than highways) by state and local government is the subsidy implicit in tax-exempt bonds.

The success of privately financed toll road projects in California has been based in part on the use of tax exempt revenue bonds to finance them. The private partners have also been given the right to raise revenues from the sale of wayleaves for the transmission by utilities of data, water and electricity. They also benefit from 'impact fees' levied from real estate developers adjoining the new road.

AUSTRALIA AND EAST ASIA

There are also some important lessons to be learnt by Europe from the experience in Australia and East Asia. The World Bank and the Australian Economic Planning Advisory Commission have recently emphasised the importance of public-private partnerships. [37] Both agreed on the need to focus on improving the quality of the infrastructure.

The World Bank considers that there is a huge need to invest in East Asia's infrastructure during the next decade as a result of under-investment in the past, continuing rapid urbanisation and the need to remain competitive. Governments in the region acknowledge that the public sector has neither the finances nor the managerial resources to meet all the emerging needs, and are therefore increasingly keen to allow the private sector to expand its role in the provision of infrastructure services. As the World Bank comments: 'While this is gradually becoming a global trend, East Asia, along with Latin America, is at the forefront of evolving the new paradigm'. [38]

The Australian commission of enquiry points to the quality of future investment. It admits that infrastructure spending has been misallocated in the past. In the future the key to achieving a proper match between infrastructure needs and supply is better project appraisal and infrastructure planning. The task force considers that a formal national infrastructure plan would be overly bureaucratic, but it recognises a need for a more coordinated approach to infrastructure planning and investment appraisal at the federal level. Good processes also involve rational pricing mechanisms, such as user charges and an appropriate division of responsibility between public and private sectors for providing infrastructure services. This framework would see funds allocated to projects with the highest social rates of return.

The major recommendation of the Australian task force has special reference to the European Union, namely that an Infrastructure Advisory Committee be established to coordinate the planning and appraisal of infrastructure investment and services that cut across state borders or are nationally significant. The Committee would advise on:

- future infrastructure needs, including the wider community effects of investments;
- viable new projects and condition of existing assets;
- infrastructure areas and national projects that deserve priority and their planning;
- best-practice benchmarks for infrastructure planning and investment appraisal.

Gaps in expectations

The World Bank emphasises that in East Asia one of the basic reasons for protracted negotiations and frustrations between public and private partners is the misunderstanding about the degree of perceived and real risks in a project. Host countries tend to perceive much lower risks than do sponsors and lenders in the private sector. Private participation is often frustrated by a lack of clarity about the government's objectives and commitments and the complex decision-making process. In most countries private participation has been seen primarily as the means to raise additional financial resources to overcome budgetary restraints rather than to improve efficiency. The governments that have clarified their objectives and given a high priority to attracting private investment, particularly Malaysia and the Philippines, have done much better than other, larger countries in the region, such as Indonesia.

The unbundling, mitigation and management of risk is one of the primary issues, and requires an appropriate policy, legal and regulatory framework, as well as mutually acceptable mechanisms for neutral arbitration procedures to enforce contractual obligations. In addition, for sponsors to assume commercial risks they must be allowed to make their own decisions on the technical and managerial aspects of the project. On the other hand, under a case-by-case approach, arguing that each project is unique, sponsors can try to negotiate mitigation of each and every risk by the state. **To avoid that and to increase transparency and competition, attempts are being made in a few East Asian countries to develop 'templates' for each major sector, clarifying ahead of bid competition who will mitigate what risks, and how.**

Stimulating public-private partnerships

Infrastructure development in Asia is increasingly seen as a core activity for many firms, as the creation of new business and not merely as an extension of construction work. Pang Chung Min, a Hong Kong banker, has expressed the view that this indicates a significant difference between Asia and Europe. [39]

Most of the private sector involvement in East Asia and the Pacific has been in telecommunications, power and toll roads where there is an assured revenue stream. As we have suggested in Chapter One, the European Union, in selecting the high speed train TENs, has given priority to the most difficult projects.

3 | Injecting Private Enterprise

A range of public-private partnerships already exists in Europe and in other regions of the world. So far, however, the repeated objective of making them an important vehicle for Trans-European Networks has not matched expectations. A recent report for the Commission points out that the large-scale TENs priority projects 'are, by their very nature, few and far between, the market opportunities open to the private sector have been very limited. There was therefore little prospect of any individual supplier being able to build up a portfolio of schemes over which to spread the sunk costs of entering a new market'. [40]

Furthermore, the transport priority projects are dominated by the high-speed train (HST), which until recently had been the province of the state railway companies. The French, German, Spanish and Italian railways have built up design expertise for major-speed lines and see themselves as the natural operators of both the infrastructure and the services. They are suspicious of entering into an arrangement which would bring private sector partners into an innovating, management role that could identify substantial cost savings at the design/build stage as well as during operation. For example, it is by no means sure that the private sector would choose the ambitious technical solutions proposed by the state railway companies. For passenger HST, the private sector may well prefer tilt-train technology instead of full 300 km per hour HST technology, thereby obviating the need for extensive infrastructure works. Similarly, for example, in the case of the Brenner axis, the private sector might settle for a high performance freight line, where characteristics such as reliability, the ability to carry very long trains and an independent US style management would count for more than pure speed.

Design-build-finance-operate

The European Commission has clearly recognised that **private partners will only accept participation in the risks entailed in TENs if they can share at least a controlling responsibility in the construction and management process.** EU legislation has already prepared the way for open access, and for the separation of track ownership and management from train operations. [41] Commissioner Kinnock has backed the concept of 'Design-Build-Finance-Operate' (DBFO) supported by substantial equity supplied by the project owners rather than at arms length by the financial markets. [42] Owner-equity backed DBFO schemes mean:

- the designers are the future operators, with quality and capacity levels optimised in a thirty year perspective;
- the designers are the builders and suppliers, bringing into the planning process precise knowledge of state-of-the-art technology;
- the builders/suppliers are the operators, having an interest in keeping costs down and completion times short;
- the builders/suppliers/operators carry out their own financial engineering which means *inter alia* keen attention to expenditure timing. For the capital markets, comfort is increased by lending, not just to a project, but indirectly to the large companies which make up the DBFO consortium. Comfort is further increased by more reliable cost and revenue forecasts carried out by the risk takers themselves, as well as cost control;
- the DBFO actor has control over its budget. It is insulated from the vagaries of annual public budget reviews, benefiting from substantial time savings.

Channel Tunnel Rail Link

The Channel Tunnel was the first attempt made in recent years to develop a major link under private concession. It has created very negative attitudes among bankers and investors about investing in large projects. However, whereas Eurotunnel, which designed, built and now operates the tunnel rail link, was problematical, the London & Continental Railway (LCR) project, which is building the English end of the rail link, is a good example of a DBFO scheme which should succeed. (See Table Five.)

As in the case of Eurotunnel, LCR's financial plans assume a major public issue of equity. Before then, equity finance is to be provided by the core shareholders. There will also be substantial debt facilities. [43] But the ways in which LCR differs from Eurotunnel are significant. Among LCR core shareholders, there is real transportation expertise. Eurotunnel's construction contract was signed with the five UK and five French contractors (TML) when they were all heavily represented on the board. With LCR, the construction contract will be subject to real competition, avoiding Eurotunnel's conflict of interest. Furthermore, the Eurotunnel/TML contract led to inevitable ambiguities as it was signed early in the project without detailed design specification. It contrasts with over six years of detailed work done by Union Railways on the Channel Tunnel Rail Link (CTRL).

Unlike the Channel Tunnel project which encountered major unforeseen engineering and safety clearance difficulties, the level of uncertainty on the costs side of CTRL is thought to be quite limited especially as the safety requirements have now been established. CTRL will acquire the already up and running Eurostar passenger services and rolling stock at present run by a British Rail/SNCF/SNCB subsidiary. CTRL will begin therefore with an existing revenue stream. Moreover, once the new terminal is opened at St Pancras, the journey from London to Brussels and Paris will be cut by about 45 minutes, giving the train a real competitive edge over air services. In addition to international passenger and freight services, the new line will also generate significant national traffic as travel time from Kent to London will be halved.

Finally, there will be a substantial contribution of government subsidy through PFI to LCR's costs. Because much of the benefit through reduced journey times and the easing of road and airport congestion, will not be recouped through increased fares, the social cost-benefit has been estimated at 8 - 10% whereas the expected financial return is significantly less at 3 - 4%. By contrast, the Channel Tunnel, on Mrs Thatcher's insistence, was funded wholly by the private sector. As Eurotunnel's chairman, Sir Alastair Morton, has recently commented:

'The Channel Tunnel was a very primitive form of pre-PFI project. Above all, it lacks in this country what it could have in France — the primary characteristic of a public authority (ministry/agency/local government) purchasing a service via a *concession de service publique* from a private sector concessionaire whose funding is a negotiated blend of public and private funds adapted to the mission. Eurotunnel has a concession but in this country it is neither a public service concession *à la française* nor a PFI *à l'anglaise*'. [44]

TABLE FIVE

Channel Tunnel Rail Link

EUROTUNNEL	LONDON & CONTINENTAL RAILWAY
Founders had no transportation expertise	Founders are companies with proven transportation expertise
Design and construction characterised by conflicts of interest	Owner/operator with contracts awarded competitively
Design specification not determined in detail at the time of contract	Over 6 years of detailed work undertaken by Union Railways
No revenues until the opening of the Channel Tunnel	Existing revenue stream
Expensive delays and additional requirements due to safety review	Safety requirements have been established
Wholly funded by the private sector	Substantial financial contribution from government and risk sharing

Source: SBC Warburg

Motorways

Public-private partnerships have been utilised to build and operate motorways since the mid-1950s in European countries that needed to develop their networks very quickly. The European Commission has pointed out that motorway concessions have proved easier to develop for various reasons:

- all users of motorways are private whereas users of the railways infrastructure are the railway companies themselves;
- road construction, maintenance and operation involves management techniques that are widespread in the private sector, whereas the technical know-how in the railways has until recently been restricted to state rail companies;

- construction and operation of a toll motorway section can be easily separated from the whole motorway network. Separation of a given railway section has been considered less conceivable although having different track and operating companies is already a fact in the UK and is under consideration in other EU member states;
- motorway traffic on major links has increased steadily in the long term with the exception of one or two periods, such as the second half of the 1970s;
- various public funding sources exist for road construction and maintenance (state, regional and local budgets, specific taxes on fuel and vehicles), whereas nothing similar exists for railway infrastructure, which can be subsidised only from general state budgets or through in-kind contributions. [45]

The House of Commons found that traditional public procurement contracts awarded on the basis of the lowest tender could achieve a false economy. Cases were cited where there were 40% over-runs in the out-turn of traditional road building contracts. In DBFO contracts, the risk for overruns would be borne by the private sector unless the risk arose from a change in public policy, such as planning delays. Furthermore, as we have noted in the previous chapter, because the private sector is being remunerated on the basis of tolls (either real or shadow), it is in the interest of the promoter to ensure that the motorway is as maintenance free as possible. For example, Cofiroute, the French private sector road concessionaire, has pioneered the phased development of highway pavements, so that pavement strength is increased as demand builds up, leading, so the company claims, to significant reductions in the lifetime costs. [46]

Airports

Public-private partnerships play an important role in airport development and operation. Traffic flows are growing continually and commercial risks over the long term are low; private partners can participate in various kinds of revenue streams such as duty free shops, restaurants and airfreight logistics. The public partners, which can be state, regional and local government, are usually favourably disposed to airport development for political, social and economic reasons.

Malpensa Airport, near Milan, is one of the fourteen TEN priority transport projects. The existing Milan airport system is expected to reach saturation point at the end of the century. The transfer of traffic from Linate near the city to a new regional hub at Malpensa located 50 km away will have environmental advantages, particularly in reducing noise pollution. The project will be financed partly from state and regional grants, partly from the cash flow of SEA, the management company of Milan Airports, and the remainder from an EIB loan with the risks shared between the European Investment Fund (EIF) and a consortium of Italian banks. The project sparked off a controversy in Brussels about whether, under current regulations, airports were eligible for EU support. The Italian government requested further EU support for feasibility studies concerning rail and road access to the airport. For a time, the Malpensa project was shuttled back and forth between the European Parliament and the Council. The case underlined the importance of establishing at the initial stage political commitment to public-private partnerships.

The bidding process

The bidding process in DBFO projects is considerably more expensive than traditional public works tenders where the state broadly specifies a project and consultants develop the details sufficiently to cater for a public tender process. The UK Treasury view has been expressed as follows:

'The traditional public sector way of doing business ... is to put a specification out in tremendous detail to the private sector to do whatever it is, build a road, build a hospital, put it out to maybe five firms and maybe get back five bids. You will get back five very similar bids because the specification was so rigid there is hardly any room for manoeuvre, so the decision at that point is very simple: 'which is the cheapest?' By comparison, when the PFI bid comes back, the specification that will have been sent out will be much more open, will specify the outputs it has sought and left it to the private sector to come back with the solutions to that, solutions which will usually have a number of variant bids in them with different degrees of risk transfer, so that instead of a simple 'the cheapest must be the best solution', you have then got a number of bids with a combination of price and risk transfer to deliver the output and you compare it with the traditional way of doing things where hopefully you have identified and costed the risk of that solution *ex ante*'. [47]

The British government regards competition as a vital element in the bidding process in order to ensure that the public sector obtains the best possible value in the purchase of the services it requires. The French government has adopted a far more informal approach. As Jean-François Poupinel of Cofiroute has pointed out, the French call upon private enterprise not in order to promote competition but to ensure that the work gets done well. [48] In France, the favoured private sector partner is identified using an informal process of negotiation, the view being that the money spent on putting together detailed bids is effectively wasted and could be better put towards the construction costs of the project.

Certainly, as we have seen, there has been considerable criticism of the PFI bidding process in the UK. The Private Finance Initiative Panel is staffed with executives who have knowledge and experience of project development, but their role is advisory and they do not get engaged in the actual deal-making. Martin Laing, the chairman of John Laing, has called for a big shake up in the way the government handles PFI projects. He wants a new department, allied to the Treasury with 'heavyweight legal and financial expertise', to negotiate all aspects of PFI contracts on behalf of spending departments. [49]

Transfer of risk

We have already noted how risk sharing is fundamental to public-private partnerships, and have suggested that risk should be allocated to the partner best able to manage it. The aim is not to encourage risk transfer for its own sake but to achieve the most cost effective solution. In general, optimum risk allocation is achieved when the partner which can manage the risk at the lowest cost will take on the responsibility. Coopers & Lybrand have explained the problem as follows:

> 'As risk is transferred to the private sector, value for money rises so long as the private sector is taking on risks with which it is familiar and which it is better able to

manage than its public sector counterparts. However, the rise does not go on indefinitely. There comes a point where the private sector may be asked to take on risks which it cannot control and which it may be less able to handle than the public sector. Although such levels of risk may be accepted they will be priced at a level which represents poor value for money for the public sector'. [50]

No hard and fast rules can be laid down for the allocation of risk. In general, the design and construction risk would be borne by the private sector, which is better placed to reduce the risk by structuring the project correctly. As purchasers under PFI specify a service and not an asset requirement, cost overruns cannot be passed on where payment levels have been agreed in advance. Greater operating risk can be transferred where the private sector partner is responsible for both the asset and the service. Demand risk may be more variable. Where the project effectively creates a monopoly, for example an estuary crossing with no convenient alternative routes, the public sector will wish to ensure that the users are not exploited by excessive charges. In general, however, the traffic risk would not normally be guaranteed by the public sector. This would be the major risk borne by the private partners although governments could relate the length of any concession to ensure acceptable minimum revenue. The public sector should be prepared to shoulder political and legislative risk, in particular where associated with planning and regulatory approval. (See Table Six.)

TABLE SIX

Risk Bearing

RISK:	PRINCIPALLY BORNE BY:
Design + Construction	Private
Commissioning + Operating	Private
Demand (or volume/usage)	Public/Private
Regulatory	Public
Project Financing	Private
Safety	Public/Private
Competing operations	Public/Private

Non-commercial risk

Major infrastructure projects are inevitably in the public domain and hence vulnerable to public policy risk, such as cancellation by a new government, failure to provide promised access routes or changes in safety and environmental standards. At a national level these non-commercial risks can be negotiated away. When several governments and regional authorities are involved in a transnational network it becomes more difficult. Bearing in mind the international experience we reviewed in Chapter Two, what can be done at the level of the European Union to minimise risk?

First, the EU could draw up a list of specified, non-commercial risks that would form a template for negotiations. Second, there should be an arbitration mechanism established at European level, bearing in mind however that long drawn out arbitration increases uncertainty. Third, the EU could reach an agreement to provide insurance cover for non-commercial risks in public sector contracts. Private partners would be able to claim if government undertakings had not been honoured. It has been suggested that a special European infrastructure insurance agency should be created for this purpose, and that to ensure that risks were pooled on a mutual basis across the Union, an element of reinsurance would also be required at member state level. [51]

Project authorities

The Essen European Council, on the recommendation of the Christophersen group, agreed that the setting up of legal entities at European level would greatly facilitate the coordination and financing of complex transnational infrastructure projects. It is generally accepted that projects designed by administrative rather than business driven procedures are unlikely to attract private risk capital. The European Commission argues, further, that intergovernmental methods of co-management are likely to lead to substantial delays, if not paralysis. It has drawn up a plan for a model project authority for cross-border infrastructure that consists of four elements:

- an agreement between the member states involved;
- a committee consisting of delegates of the member state governments;
- project promoter;
- project company which acts as infrastructure manager. [52]

In order to establish a genuine partnership and to take advantage of private sector skills at an early stage, the project's conceptual definition should be undertaken either by a public-private partnership or by a private entity with a direct economic interest in the success of the scheme on completion. On the public side this may involve state, regional and local authorities. On the private side, the services of a design company or consultancy would usually be employed by the shareholders to exploit to the full all the technological possibilities and low cost options.

The European Economic Interest Group (EEIG) provides a valuable instrument for carrying out the preliminary technical and economic feasibility studies; but EEIGs do not offer the limited liability required for the implementation phase of the project. Embryonic single agent structures have been created in the form of EEIGs by the respective railway companies to study the tunnel projects of Lyon-Turin and Barcelona-Perpignan. For the Brenner, studies are being undertaken by a three railway working group. Ideally, however, a wholly new company should be created to own the project. This medium-sized promoter company would assume the task of defining the project, developing the design and corresponding technical specifications, assessing the costs and expected revenues, preparing the legal work and carrying out the environmental studies. **Such companies need the protection of European Union company law, and the Commission's current attempts to revive its proposals for a European company statute must be welcomed in this regard**.

Who decides on the composition of the initial promoter company and selects the private sector participants? The European Commission suggests that the promoter company would be selected by a steering committee of representatives of the member states concerned. Consortia would be invited to present their ideas and, after pre-selection, to present initial cost estimates, including the need for government subsidies. The willingness of participants to invest their own money in the build-and-operate phase would be a de facto selection criterion. Conventional tendering procedures, with their legal safeguards, would be difficult to apply at this stage because the precise parameters of the project would not have been defined at that point. But *functional* requirements, as opposed to technical specifications, could and should be laid down by the project steering committee for the purposes of the initial competition.

When the promoter company has completed the feasibility studies, optimised the design, negotiated the rights and obligations of the public sector and outlined a financing solution, it is succeeded by a fully-fledged build-and-operate company. ECIS has suggested that for projects of limited geographic scope, such as airports, this changeover could occur without further tendering. Control against excess profitability could be exercised through profit-sharing via public equity, which eventually should help to stimulate a wider capital market for infrastructure in which builders could sell their equity stake after successful completion. For larger projects, a second competition has to take place before the construction phase. The promoter company should be allowed to tender and will be well placed to win the competition. However, other tenderers must be allowed to offer alternatives based on their own projections made with the knowledge of the original design work.

Terms of competition

There is a considerable convergence in the above approach (which is a variant of the French model) and the PFI method involved in the Channel Tunnel Rail Link. We have already noted how the British Treasury has emphasised that PFI specifications are much more open-ended than traditional tendering procedures.

In the CTRL project bidders were required to submit reference bids according to tender document specifications in respect of route alignments, time of construction and key operating parameters such as the number of peak period domestic train paths. Variant bids could be submitted in respect of detailed design, and operational and financial aspects relating to the allocation of risk between the promoter company and the government. Although the competition document is not entirely transparent, it appears that bids will be selected according to the best financial terms plus certain quality thresholds relating to technical aspects and robustness of financial and management structure. 'Best financial terms' will be judged by reference to the level and timing of government contributions required and by the extent to which the bidder is willing to take on risk. Four consortia have been invited to tender and the government announced that it will contribute 33% towards the costs of unsuccessful bids up to a ceiling of £1.5 m, subject to the bid being of a quality commensurate with the nature of the project.

London and Continental Railways have already been awarded the contract to become the owner/operator of the Channel Rail link. As we have already noted their objective is to become a major publicly-quoted transportation company. The skills of the shareholders range from project design and management of high speed trains and terminal operations to marketing and brand management. The shareholders have no interest in construction or supply contracts which will be awarded on a competitive basis.

The CTRL project, which may yet serve as a template for a major high speed TEN, is currently being vetted by the European Commission. This is being done according to existing rules on competition, state aid and public procurement but also against the background of the Essen Council which invited the Commission to encourage public-private partnerships. Neil Kinnock has committed the Commission to doing everything possible to take a final decision on exemption from the competition rules within a maximum of six months from the notification of an agreement relating to TENs. With regard to the right of access to infrastructure, the following criteria will be applied:

- all EU undertakings should be given the chance to apply for reserved capacity;
- the capacity to be reserved should be in proportion to financial commitments and in line with planned operational requirements;
- some capacity should remain to allow competing services;
- rights will be reallocated if they are not used;
- capacity reservation agreements should not go beyond a reasonable period.

The Commission has decided that existing public procurement legislation is a sufficient framework for the participation of the private sector in the award of concessions, as long as effective competition is allowed in the tender phase; and that such practice will ensure value for money. Commissioner Kinnock has emphasised that EC law does allow pre-tender technical discussions. (See Table Seven.)

Channel Tunnel Rail Link

THE GOVERNMENT

Objectives
Construction and operation of the CTRL by the private sector
Privatise European Passenger Services and Union Railways
Regeneration of the East Thames corridor
Minimise the risk to the government
Minimise the public sector contribution required

Contribution
Capital grant payable towards the cost of developing the CTRL
Donation of property and net assets of European Passenger Services
Domestic capacity charge for domestic train operators

Risk sharing
Changes in law or taxation
Changes in safety requirements
Government imposed changes to the CTRL
Hostilities and civil disorder

LONDON AND CONTINENTAL RAILWAYS

Objective
Establish a major transportation company

Philosophy
Owner/operator philosophy
No interest in obtaining construction or supply contracts
Construction and supply contracts will be awarded on a competitive basis

Shareholders
Shareholders bring necessary skills for all stages of the development of the new company:

Bechtel	Infrastructure design and project management
Halcrow	Transportation planning and tunnel design
London Electricity	Electrical infrastructure design and operation
National Express	Marketing, franchising and terminal operations
Ove Arup	Engineering and design
SBC Warburg	Financial advice and structuring
Soférail	Expertise in high speed rail link construction and operation
Virgin	Marketing and brand management

Source: SBC Warburg

The way ahead

There is an increasing realisation in Europe that public-private partnerships have a vital role to play in improving the quality of infrastructure. Considerable problems have had to be surmounted which have not been made easier by the priority TENs being dominated by very large projects of which the largest are high speed train projects. The public sector state railways are gradually being persuaded that private partners have an important role to play, not least in the provision of finance.

The biggest revolution is the input of the private sector partners to the design process. In this connection the UK's Private Finance Initiative has been innovatory. France has pioneered the involvement of the private sector in the building and running of concessions — for example, by exploiting the enterprise of private water companies, some of whom have now become world leaders. **The pragmatic French approach is making an important contribution to the developing European model of public-private partnerships, although EU practice favours a greater openness in awarding concessions without restricting choice.**

The European Union has the possibility of drawing on the best experience of public-private partnerships in its member states and other regions of the world. **The Channel Tunnel Rail Link is demonstrating how the concept of Design-Build-Finance-Operate can work. But in order to make more progress, companies need to be able to build up a portfolio of schemes in order to spread the costs of entering a new market.**

4 | The Role of Private Finance

Interest in public-private partnerships by governments at successive European Councils is largely based on the acknowledgement that public budgets cannot suffice for trans-European networks. The Delors White Paper quite rightly assumed that the member states had virtually no margin to increase public financing for TENs. The danger is that public-private partnerships are seen as the deliverers of financial engineering that will inevitably bridge the investment gap. **In the energy and telecoms sectors, where the role of private finance is already established and growing in importance as a result of de-regulation, competition and privatisation, the lack of public finance may not matter. The case of transport TENs is more problematical.**

Mismatches

A characteristic of many transport projects is the long period that elapses before there is any revenue to meet financing charges. First, there is the planning and preparation phase where there are high risks of delays and cost overruns, when unpredicted environmental or regulatory problems can occur, which might result in fundamental redesign of the project making it less viable than originally assumed. This phase of the project typically requires public sector grants and/or financing by high risk venture capital.

The second phase of a major transport project is the lengthy construction period, where there is considerable risk of further delay and cost overrun especially if the project is insufficiently planned and requires major modifications as it develops. The construction phase can largely be financed with bank loans. The risk of cost overruns or delays should be covered by risk capital where possible in order to ensure that the debt service of the project does not deteriorate if these risks materialise.

When the project is completed there can still be a problem of revenue shortfalls in the initial operating phase. These can arise through consumer resistance to pay for the new service or time lags and inadequate marketing, resulting in the under-utilisation of capacity. On the other hand, there might be 'upside' chances of higher revenues, in which case it might be possible to refinance the high cost construction loans with improved terms. The balance of the financing needs should ideally be financed in this phase not by debt but with equity or grants.

Even in the mature operating phase, when cash flow is established and gradually strengthening with increased traffic volume and higher charges, risks can remain. Floating rate debt in a period of rising interest rates can be expensive if revenues are fixed. So construction finance has to be refinanced with long-term and low-cost senior debt. Projects in this phase have been found to be attractive to institutional investors in North America where funds are seeking long-term assets with inflation adjusted revenues. Other funding instruments in the USA could be provided by revenue bonds, fixed term debt or ordinary equity. In the mature phase equity or subordinated loans will still be required to provide a buffer or hedge against temporary cash flow deficits.

A major constraint for TENs lies in the limited availability of risk capital, in particular venture capital and equity for long-term infrastructure investments. Private sector venture capital investors, to the extent that they exist, typically prefer investments with 20 - 30% initial return in the development phase. In Europe, some large building and engineering groups have been accustomed to carry out design studies and construction work. Their normal inclination would be to work on contract to a project promoter and take as little of the risk relating to the project as possible. In the energy, telecoms and water sectors, large operating companies are willing to provide risk capital for new investments, whereas in transport projects, contractors or main suppliers are often expected to provide equity financing. Contractor equity is a potential source of problems. In the construction phase there is a natural conflict of interest between the long-term owner and the executing contractor. Most contractors, particularly in the UK, are weakly capitalised and have very limited scope for long-term equity investment. To the extent that they have been obliged to supply risk capital themselves in the absence of alternative investors, it has come at a high cost. [53]

Grants

At present grants are extensively used by the EU and member states to cover 'equity gaps' in transport projects. From the government's viewpoint, grants are the most expensive form of public support as no return of cash flow will ever be realised. This characteristic provides the grant with maximum leverage effect and has been used extensively in the USA for this purpose. Grant financing also enables the public sector to conduct hands-off management. It can also be used indirectly to subsidise consumption by providing low cost access to transportation projects. The problem is that it is difficult to avoid excessive subsidising as grant decisions have to be made early in the project cycle at a time when it is difficult to quantify project risks accurately. Construction and operating costs and debt servicing might turn out to be lower, and revenues higher, than foreseen. In this case excessive grant financing might actually be a disincentive to make projects financially viable and to mobilise private capital. [54]

It has been suggested that a contingent support mechanism could require the public agency to pay out more in grant to the project promoter if costs were higher or revenues lower than expected — and less if events turn out favourably. Because the private sector is likely to attach relatively more weight to adverse outcomes than the public agency, the overall package could be capable of being seen to be more attractive to both parties than a fixed price bid. [55]

Public equity

There is a case for substituting public sector grant financing by risk capital with equity features. Since *public* stake-holders invest equity in expectation of 'external benefits' to the economy in general, their return might not initially comprise financial dividends. In the early years of a project the net profit would be available for the *private* equity holders, while the inherent risk reduction properties of the scheme should lead to affordable debt charges. Once the project reaches an agreed threshold of profitability, the public equity stake could benefit from a pre-determined profit-sharing mechanism, and receive dividends. [56]

Public equity financing should be subordinated in order not to increase debt service obligations. As far as possible it should also be a passive investment that cannot lead to excessive public sector interference with the project management. The object would be to attract pension fund and insurance company investment. At the same time a public equity holding can ensure that any upside revenue or capital gains over and above the expected rate of return is captured for the community as well as the private shareholders.

Tax relief

Very large sums of money will be needed to build TENs from new or to link and upgrade existing networks. Clearly, the higher the rates of tax depreciation available and the higher the proportion of capital spend eligible for such tax depreciation, the lower the amount of finance needed to construct the TENs. Rates of tax depreciation vary from country to country but typically, tax depreciation for infrastructure projects takes place over several years (up to fifty in some cases). Generally, the shorter the expected useful life of the asset, the faster it is depreciated for tax purposes. Accelerated rates of tax depreciation on TENs projects should have the effect of increasing investment in TENs. Given the time limit for use of brought forward tax losses in many European countries and the cashflow profile of TENs projects, the project vehicle may wish to lease plant and other assets from companies which can use tax depreciation allowances in the short term.

Capital and revenue subsidies are normally taxable. However, investment in TENs would be much more attractive if capital subsidies were regarded as some non-taxable form of capital contribution, as public equity or quasi equity, and not as reducing the cost of assets for tax depreciation purposes. Investment would also be encouraged if revenue subsidies were not treated as taxable income. There is some precedent for this treatment: in the UK, regional development grants are not treated as taxable.

The project vehicle will be keen to get relief as soon as possible for pre-operating costs. In some cases, an existing operator in the sector would expect to get immediate offset of pre-operating financing and other costs against profits of its existing operations. On the other hand, new entrants to the sector will not obtain tax relief for the costs incurred until the project starts operation or possibly until it breaks even. If the vehicle is a partnership, which is tax transparent with income and expenses being treated as those of the partners, it may be possible for the partner to set pre-operating losses against profits from other operations. However, this route will only be viable if financial and other commercial needs are also met. A tax transparent structure has been used by the French, Dutch and Belgian state railways operating the Trans-European Express.

If the Commission were to establish a level playing field for existing operators and new entrants, it could seek to persuade member states to treat pre-operating costs as trading expenses relievable when incurred. However, given past experience on harmonisation of tax treatment, this suggestion must be unlikely to be effected in the short term.

Venture capital

Ranjit Mathrani, chairman of Vanguard Capital which specialises in the finance of major infrastructure projects, has emphasised the need for special purpose development vehicles, dedicated capital funds and specialist infrastructure companies. [57] He considers that in the pre-construction phase about 80 - 100% of the private equity will have to be provided by the promoter, although he doubts whether the sophisticated deal-making skills that are required flourish in many companies.

In the UK, the Innisfree PFI fund was launched in January 1996 as a private equity fund to invest in projects promoted under the government's Private Finance Initiative. Hermes Investment Management which manages Britain's post and telecoms pension funds and the AMP Asset Management which looks after investments for the Australian Mutual Provident have each invested £25 m. The AMP Society manages the Develop Australia Fund which has a capital base of around A$ 150 m. Part of these funds is invested in infrastructure. The AMP Society has also announced the formation of a A$1 bn fund to channel superannuation and life insurance savings into infrastructure. [58]

Innisfree has identified 45 PFI projects with a total cost of £4 bn as potential investment vehicles. As a first step it is looking at two transport schemes, a hospital and an office scheme with a combined value of £300 m. The chief executive, David Metter, told the *Financial Times* that private investors, normally operators and promoters of PFI schemes, would typically be expected to provide 10 - 15% of the total cost of projects. Innisfree would expect to provide about a quarter of this equity. The bulk of finance would continue to be provided by borrowing. The Fund would expect to generate internal rates of return of 20 - 30% on sale of investments once projects were operating and generating revenues. [59]

The challenge is to encourage the growth of both equity funds, such as Innisfree, that will invest in the pre-operational phase, and longer term investment funds that could attract pension funds into infrastructure projects once the revenue stream has been established. **The European Union could play an important role in providing equity or quasi-equity financing rather than grant finance**. The existence of funds that could provide liquidity for sponsors' equity would strengthen the financing of public-private partnerships.

European Investment Bank

The bulk of the financing in the construction and operating phases of infrastructure projects at present comes from long-term debt. The European Investment Bank (EIB) is the most important single source for financing TENs. The Delors White Paper estimated that Ecu 2 bn would be required annually on top of existing public funding by the member states. The EIB was to provide Ecu 7 bn with another Ecu 8 bn raised mainly from the financial markets through Union Bonds. When the Union Bonds were blocked by the Essen European Council, it was agreed that EIB lending to priority TENs should be further increased through the establishment of a special 'window' through which the Bank is expected to provide up to one third of its lending in future years. The 'window' has the following features:

- loans of extended maturity, and grace periods to match debt repayment and cash flow;
- refinancing facilities for the commercial banks;
- partnerships with the capital markets;
- involvement of the EIB in conjunction with the European Investment Fund (EIF), in the earliest possible stages of the financial and contractual structuring of a project to facilitate suitable arrangements.

The EIB has so far approved financing of Ecu 6.3 bn for nine of the fourteen priority transport TENs. Finance contracts of Ecu 4.6 bn have been signed for projects covering all or part of:

- Italian section (Verona - Brenner) of the Brenner high-speed railway line;
- TGV Paris - Brussels - Cologne - Amsterdam - London (PBKAL);
- principal rail link in Ireland between Cork, Dublin and Belfast;
- road and ferry link between Ireland, the United Kingdom and Benelux;
- many sections of motorways in Greece, Portugal and Spain;
- sections of road and railways links in Finland forming an integral part of the 'Nordic Triangle' linking the capitals of the Scandinavian countries;
- Øresund fixed rail and road link between Denmark and Sweden;
- extension of Malpensa airport outside Milan.

The EIB has considerable experience over a period of years of promoting private sector financing of infrastructure. Its president, Sir Brian Unwin, gave the following examples:

> *Directly* through loans to the private sector and by wholesale lending through the banking system. At the end of 1994, almost half of the Bank's outstanding loans were covered by some form of private sector guarantee compared with 10% in 1985.

> *Indirectly* by making it easier for a promoter to access private debt and equity in its own right. The capital market as well as the commercial banks are more willing to provide project finance when the risks are shared with the EIB. The Bank has a reputation as a quality 'AAA' borrower and applies equally high lending standards through rigorous criteria of financial, economic and technical appraisal. [60]

TABLE EIGHT

European Investment Bank: Additional Finance for Priority TENs

1. FINANCING OF INTEREST DURING CONSTRUCTION

The EIB finances interest during construction as part of project costs. The debt service in the early stage of a TEN can be reduced through the capitalisation of interest to be repaid over the life of the loan.

2. PROVISION OF LONG MATURITIES

By providing loan maturities in excess of 20 years in certain projects, the EIB would help TENs to minimise the amount of project cash flow which has to be devoted to debt repayments in the early years of a project.

3. CO-FINANCING OF PROJECT DEBT

Many commercial banks are prepared to provide construction finance but do not wish to be tied into a project and take revenue risk over a long period. The EIB is prepared to consider arrangements to take the banks out of the project when complete under a framework agreement put in place at the initial financing stage.

4. EXTENDED GRACE PERIODS FOR CAPITAL REPAYMENT

TENs projects can have a slow build-up of positive cash flow after operations begin. The EIB can provide 'bullet' loans where capital is repaid in one lump sum at the end of the life of a loan.

5. FIXING LOAN RATES IN ADVANCE OF DRAW DOWN

Advance funding enables promoters to protect themselves against any increase that may occur between the establishment of borrowing facilities and the time they are drawn upon to finance construction

6. FRAMEWORK CREDIT AGREEMENTS

In certain projects the EIB would be prepared to enter into a framework credit agreement under which they will undertake to provide a substantial part of the finance required, subject to commitments made by the promoter. Disbursements under these agreements are made at open rate contracts which give the promoter the possibility without commitment fees to draw upon an agreed line of credit at the rate of interest prevailing on capital markets at the time (as distinct from the time of the initial commitment).

Source: Gérard Patrice, *The Role of the EIB in promoting private sector finance*, Workshop on Private Sector Finance of Community Interest Projects, DG II, European Commission, Brussels, June 1994.

Finally, the EIB is prepared to participate at the earliest possible stage in working parties for priority projects. The aim is to minimise the financing costs and risk, and to ensure the appropriate allocation of risk and reward to the promoter, governments and the various financiers. Such an approach has already been tested to varying degrees in the case of several large transport projects such as the second Severn Bridge in the UK, Wijkertunnel in the Netherlands, the East-West gas interconnector in Germany and the Tagus Bridge in Portugal. Currently the EIB is involved in several TENs working parties such as the TGV-Est linking Paris-Strasbourg-Luxembourg-Frankfurt and the Belgian section of PBKAL. In some cases, the EIB is part-financing with the European Commission the feasibility studies for these large projects.

The EIB has become the most important and attractive source for financing TENs. There are constraints on its lending, however, imposed by its requirements for adequate guarantees. The EIB has never had a major default or credit loss and safeguards strictly its triple AAA status. One element is a very cautious credit risk policy. For private sector infrastructure projects, the bank would normally require public sector or commercial bank guarantees which add to the financing costs and are constrained by the banking sector's capacity and its risk taking willingness. (See Table Eight.)

European Investment Fund

The EIF was created uniquely and specifically in 1992 to help finance TENs throughout the Union. The EIF has two big shareholders, the EIB with 40% of the authorised capital and the European Commission with 30%. The remainder of the shareholding is subscribed by over 70 public and private banks. The total authorised capital is Ecu 2 bn and its principal function is to offer financial guarantees. The EIF is allowed to guarantee up to three times subscribed capital which at present would equal Ecu 5.3 bn. It has to operate on a commercial basis and pay dividends and accordingly has to spread its risk and limit its exposure to any one project.

Its primary role is to carry risks that the private sector is unable to bear but for this service it requires a fully commercial rate of return. There are three areas where the EIF can bring particular added value by guaranteeing:

- very long maturities (over 20 years);
- loans during construction or pre-operational period;
- subordinated debt or quasi-equity.

The EIF can also contribute equity to TENs operations and to small and medium sized enterprises but the total funds available are small (perhaps Ecu 75 m over the next three years). So it may make sense to support quasi-equity for a few projects rather than spreading it thinly across several large projects. Projects in the pipeline with EIF guarantees include DBFO roads in the UK, the Channel Tunnel Rail Link, additional mobile telephony and data networks as well as other projects in the gas sector. The Fund's involvement in supporting the High Speed Train network in Europe will depend on whether the financing is made in the traditional way — that is, on the balance sheets of the public sector companies — or on a project finance basis with limited or no recourse to governments. The more projects are developed 'off-balance-sheet'; the greater contribution the EIF can bring, by providing the support necessary to attract private finance on acceptable conditions. [61]

Direct market access

At present commercial banks are the second largest providers in EU countries for infrastructure projects. These banks traditionally prefer short and medium-term lending with maturities of up to 5 - 10 years in order to match their lending portfolios with their deposit structures. By contrast in the US, large infrastructure projects are mainly funded directly from the capital market through revenue bonds with maturities of up to 30 - 40 years. The demand for long-term securities is based on:

- the need of institutional investors, for example, insurance companies and pension funds, to match their long-term liabilities;
- the need of funds investing long-term savings to acquire stable revenues;
- the interest of institutional investors in using the price volatility of long-term bonds for portfolio management.

There are several reasons for the situation in Europe. One is the lack of a fully integrated capital market, and in the absence of a single currency the existence of currency risk which constrains cross-border borrowing within the EU. [62] Second, there are too few institutional investors as most countries, apart from the UK, Ireland and the Netherlands, do not have a large funded pension sector.

Third, governments have tended to regard the long-term bond market as being principally available for the funding of their own borrowing requirement. This situation may well change as a result of EMU. Public sector budgetary constraints are likely to lead to the emergence of private pension funds in most EU countries. [63] **At the same time, the single currency will accelerate the integration of the various capital markets and the emergence of a Euro long-term bond market.** The time could be ripe for the European Union to encourage these trends, which in the United States are supported by tax relief provisions for revenue bonds.

Financing constraints

Sir Brian Unwin has pointed out that: 'Although the financial requirements of infrastructure projects in absolute terms are large, the EIB continues to believe that the main bottlenecks to investment are often the highly complex techno-administrative arrangements that have to be agreed before a project can start. Constraints are often particularly pronounced in the case of large scale cross-border projects like most of the TENs, where more than one government or authority are involved'. [64]

The institutional implications of this problem are examined in Chapter Six. There is, however, one vital issue that impinges on the role of private investment. As we noted in Chapter One, if infrastructure projects were selected on the basis of meeting the required rate of financial profitability, most transport TENs would not surmount the required hurdle. It is generally accepted that infrastructure projects have to be tested against a desired economic rate of return that is usually measured on the basis of cost-benefit analysis that takes account of benefits to society as a whole, such as environmental improvement. Where the economic justification provides the case for a project going ahead, the private partner would expect that the public sector would provide grants or quasi-equity which would bridge the gap between normal financial expectations and economic benefits. **Public subsidies, therefore, have a vital role in making transport TENs acceptable to the private sector.**

The Lyon Periphérique Nord provides an example. The Communauté urbaine de Lyon (COURLY), composed of 50 local communities with a population of 1.15 m, considered that an economic return of 9.5% (and 13% if indirect benefits are included) justified the roads project despite the fact that the financial return of 5.5% would not permit the scheme to go ahead. COURLY decided to provide a subsidy of 30% of the project costs totalling around Ecu 700 m. The effect was to raise the financial return to between 7.4% and 8.8%. The scheme was now able to go ahead with the EIB, other bank loans, and EIF guaranteed subordinated bonds providing 58% of the required capital, with the remaining 12% equity from private partners. According to the Commission the return on equity (including the convertible bond) will now be around 20% after tax — enough to persuade the private partners to go ahead and assume the construction and the revenue risk. [65]

Cross-border projects

Cross-border projects are more difficult to evaluate. ECIS has pointed out that with TENs each jurisdiction is principally responsible for the funding of the respective national section of the project. [66] **Without a deliberate effort at central coordination, economic appraisal of a cross-border project takes the form of a series of separate, uncoordinated national evaluations**. Analysis of the economic return from a London-Paris HST link would tend to examine the benefits to passengers on domestic trips — for example, London - Ashford on the CTRL and Paris - Lille in the case of TGV Nord. The effects of decongestion on other rail routes, roads and airports and environmental benefits would also be included. The other stream of benefits analysed would be the national share of the international economic benefits. In both the UK and France, however, the rules governing the determination of state subsidy explicitly discount the economic benefits to non-residents. The British evaluation would not take into account the estimated benefits to French and other international passengers arising from the UK's investment, and vice versa.

The case for Trans-European Networks is that the completion of the 'missing links' between the national transport, IT and energy systems will benefit the European Union as a whole. **The Union's institutions therefore have a major interest in identifying and securing the supranational advantages arising from the TENs.** The present method of calculating economic return is understandable only in relation to justifying any extra burden that will be placed on the taxpayer of individual member states. By excluding international economic benefits, member state evaluations fail to take account of a substantial part of the 'consumer surplus' that accrues from TENs.

ECIS and the PBKAL working group chaired by the European Commission have both recalculated the economic return to take account of this defect, and have found that if the full international benefits are included the project's rate of return increases from 7.2% to 9.5% — thereby passing the internationally recognised hurdle of 8%. The EU component of the economic return at 24.2% is of a very high order, reflecting the significant international character (five countries) of the PBKAL project. Using EU guidelines, the project would then qualify for a subsidy up to the TENs funding limit of 10% of the investment cost laid down by Ecofin. This would imply additional EU funding of the order of Ecu 1 bn.

User charges

Transport projects may not pass a required financial return because user charges are either too low or non-existent. Citizens have been used to thinking of transport as a public good. This is particularly the case in road transport where users are generally not required to pay directly for their use. Road pricing schemes are now under active consideration. In the UK, fuel and vehicle excise duties amounted to £16 bn in 1993-94, while £6 bn was being spent on road maintenance, construction, enforcement and administration. However, should indirect costs be taken into account (congestion costs £13.5 bn, pollution £2.8 bn, accidents £7.5 bn and noise £0.6 bn) road users are seen to be heavily subsidised by the rest of society. [67]

In December 1995 the Commission published a Green Paper entitled *Towards Fair and Efficient Pricing in Transport*. [68] It suggests that the only way to curb congestion in the long run is to set an explicit price for infrastructure capacity. The introduction of congestion charging would bolster the efficient provision of infrastructure; without it, the tendency would be to build more infrastructure than may be necessary. Studies in the US have indicated that moving towards congestion pricing could lead to annual cost savings of $7.75 bn or nearly 18% of total highway expenditure.

Congestion charging would also raise significant revenues which would go towards the recovery of the capital costs of the network. Such an approach would have several advantages. The revenues from such charges would remain within the transport sector and would thereby benefit those who pay them; they might allow other taxes to be cut; or they could be used to finance other parts of a comprehensive traffic strategy, such as route guidance systems or public transport provision. **In the long term, congestion charging should greatly increase the efficiency and equity of transport**.

The Commission Green Paper recognises that **such a reform of transport taxation would add to the financial viability of public-private partnerships by providing a stable revenue source**. Income from user charges should not only compensate for marginal costs but also cover repayment of the initial capital outlay. This points up the importance of the public sector element of public-private partnerships in transport TENs. A purely private funded project would be tempted to maximise revenue to the extent that capacity utilisation would be driven below the optimal rate from the point of view of regional development. Fear of such an outcome has informed the development of the Øresund fixed link.

Øresund Fixed Link

It is estimated that the completion of the first fixed link between the Scandinavia and mainland Europe will lead to an increase in passenger rail traffic of 450%. The ambitious fixed rail and motorway link across Øresund will extend about 16 km between the Danish coast south of Copenhagen and Lernacken on the Swedish coast. It is a joint venture of Danish and Swedish state owned public limited companies costing about Ecu 3366 m and expected to be opened in the year 2000. The fixed link is being financed by capital market loans guaranteed by the Danish and Swedish states. The two governments donated seed capital of Ecu 6.8 m. and Ecu 21 m. has been provided by the EU TENs budget, and further funds are being sought for access roads and feasibility studies.

The Commission has asked the consortium to investigate the possibility of introducing private sector financing on a substantial scale. However, little attempt has been make to do so as there is little or no history of private sector involvement in transport infrastructure, particularly in Denmark. Both governments appear to believe that the project will be profitable and that any upside revenue should not be given to the private sector. Public sector ownership is regarded as essential in order to minimise potential conflicts between private and social interests, particularly on environmental issues. The 'user pays' principle will not only apply to the road link but will also be used for the railway line. Danish and Swedish railways will pay a yearly flat rate to the consortium for the use of the infrastructure. However, it is expected that about 80% of the fixed link revenue will come from passenger and freight road traffic, as tolls will be based on the equivalent Helsingor-Helsingborg ferry crossing which has a monopoly of surface traffic until the new link is opened in 2000.

Shadow tolls

Experiments are underway to introduce systems of electronic charging with smart cards that would simultaneously give the motorist information on traffic conditions and weather reports. In the meantime, the UK has introduced a system of financing DBFO road projects through 'shadow' tolls. Their advantages over 'real' tolls are that they cause no diversion of traffic on to minor roads, traffic levels are not distorted, and revenue begins during construction and increases as road is completed. The disadvantages of shadow tolls are that no actual income is raised to pay for infrastructure, and they do not enable price to be used as a management tool.

Shadow tolls are designed to ensure that motorway projects can secure a high level of fixed interest borrowing and a relatively low level of equity financing. An example is presented by the Road Management Group that has secured two contracts to build DBFO roads in the UK. Road Management Group comprises Amec and Alfred McAlpine of the UK, Brown and Root of the US and Dragados, a Spanish road builder; it will assume the risk of construction and maintenance; the traffic risk will be shared with the government through shadow tolls which have made it possible to finance the project principally by a 25 year £165 m Eurobond offering in the capital market rather than by a syndicated floating rate debt. The EIB will provide £111 m of further loan finance and its facility and the bond issue will be credit-enhanced through guarantees of the EIF.

Need for more money

Transport TENs, as we have seen, are characterised by mismatches between the demands of the project and the supply of finance. The EIB has done stalwart work to provide loan finance with an overall commitment to priority projects since 1993 of Ecu 5.8 bn. The creation of the TENs 'window' has gone a long way to meet the unique requirement of certain projects through lengthening maturities over a 20 - 30 year period together with extended grace periods.

The problem remains, however, that several projects will only pass the internationally acceptable rate of return if they attract substantial non-debt finance as well. So far EU member states have not been willing to follow through the Essen Council's 'informal' decision to provide another Ecu 1 bn of grant-aid to TENs. The fact is that there is a stronger case for public investment of equity rather than grants. The European Investment Fund has the possibility of making equity available for TENs, but the amounts are ludicrously small in relation to need: they should be increased. **Europe needs an EIF that is actively involved in taking equity stakes in TENs. Such an innovation would give a breadth and liquidity to infrastructure funds in the European capital markets that they will otherwise take many years to acquire.** EIF stakes could be sold off when the projects had reached an acceptable revenue stream. And large industrial groups would likewise be more attracted to TENs if there were more of a prospect of selling their investments on.

The World Bank and the International Finance Corporation have been instrumental in creating $1 bn infrastructure funds in Asia. A new fund to provide finance for investment projects in Latin America has just been created. The fund which is aiming to attract $1 bn from institutional investors is being put together by former senior executives at the World Bank. The European capital market will only provide similar investment vehicles for public-private partnerships if there is a significant growth of funded pension schemes. So far **the investment of pension funds in TENs is hampered by the failure of the EU to liberalise regulations governing pension fund investment management**. There are the beginnings of private infrastructure investment funds, such as Innisfree, that are attracting pension fund interest, but the pace is slow.

The reinforced integration of the European capital markets has enormous potential both for the increased mobility of labour and for increased levels of financial investment. It has been estimated that EU pension funds could grow nine times to $12.7 trillion by the year 2020 if all national restrictions on the way assets were invested were removed. [69]

Finally there are the technical and administrative problems which characterise many transnational infrastructure projects. **Only the European Union can create the common standards and evaluation criteria which lie at the heart of assessment of TENs.** We examine how it might do this in Chapter Six.

5 | Infrastructure and Enlargement

The need for infrastructure investment in Central and Eastern Europe is self-evident, following decades of neglect, bad management and contempt for the environment. A mixture of private investment and public funding is necessarily the means to meeting critical infrastructure demands and of introducing up-to-date technologies. In one sense many of the Central and East European states have reacted more favourably to the concept of public-private partnerships than some EU member states, with their well-established public sector facilities. Many obstacles still have to be overcome, however, before public-private partnerships can make a significant contribution in Central and Eastern Europe. In particular, the legal environment must be clear and must support private participation by providing contractual protection and security.

The European Bank for Reconstruction and Development (EBRD) is at the forefront of building up public-private partnerships in the ex-communist states. It has been difficult to know what to keep from the old regime and what to discard. EBRD funded projects had reached Ecu 935 m by the end of 1995 in the whole region. The Bank is also playing a leading role in producing appropriate legislation through round tables and symposiums which will 'lay the foundations for large-scale mobilisation of private investment while striking a proper balance between public and private interests'. [70] The EBRD has tried to encourage a limited entry of new private providers without a significant unbundling of existing public utilities. Public-private partnerships are working best in areas such as power plants, cellular telephone networks, toll roads, municipal services, ports and airports.

Telecom networks

A high proportion of the privately financed infrastructure schemes in Central and Eastern Europe is in telecoms, particularly the provision of local networks with the objective of broadening access for small business and residential subscribers. Hungary, in particular, has been able to attract foreign investors because of the relative clarity of its regulatory environment. Telecoms projects have enjoyed the following advantages over other sectors:

- massive pent-up demand for service;
- relatively small initial capital requirement;
- potential for invoicing foreign-owned companies in hard currency;
- large number of potential sponsors looking for overseas equity investment;
- easier regulatory problems. [71]

In some cases telecoms projects have attracted commercial debt finance on a project basis without corporate parent guarantees. In other cases, sponsors have been willing to provide 100% equity finance or have co-financed equity with EBRD finance.

Energy and the environment

Pipelines has traditionally been an attractive area for private finance. A number of oil and gas pipeline schemes are under discussion, some promoted by international oil and gas majors, some by national utilities. In Slovenia, a company owned by the Dondi Group of Italy, will build, own and operate natural gas distribution networks that will be transferred to six Slovene municipalities after 30 years. By awarding concessions, the municipalities have been able to free limited financial and managerial resources for other purposes.

Significant progress has been made on developing privately financed power projects. Here the economic rationale varies from country to country. In Hungary there is considerable interest in developing gas-fired generation in place of coal and oil. In the Czech Republic and Poland the aim is usually to replace environmentally unfriendly capacity with gas-fired or clean coal technology. Such schemes envisage the export of at least some of the capacity to Western Europe. Many of the schemes have a district heating component where steam will be sold to municipal district heating companies to replace production by heavily polluting boilers in urban centres.

Those states that have signed Association Agreements with the European Union face the need for large capital investments to meet environmental accession standards. Few municipalities in these countries, which include Poland, the Czech and Slovak Republics, Hungary, Romania, Bulgaria, Slovenia and the three Baltic States, come close to meeting EU environmental standards in drinking water quality, waste water treatment and solid waste disposal. The EBRD has pioneered a way of streamlining the financing of eligible projects and of requiring the sponsor to make full financial and economic disclosures. One example is an agreement with Lyonnais des Eaux where the EBRD will provide equity and loan finance for 10 to 15 small and medium-sized investments to supply municipal services including water supply, waste-water treatment, district heating and solid waste management. The Bank will provide Ecu 70 m supporting a total investment of Ecu 234 m. As the EBRD has commented: 'This constitutes the first-ever financing by an international financial institution of private investments in municipal services and infrastructure in the region. It offers an innovative and effective vehicle to support private sector provision of these services which is expected to bring major benefits'. [72]

Transport

The European Union has given a high priority to improving network links with Central and Eastern Europe. Although substantial sums have been earmarked by the European Investment Bank, the EBRD and the EU's Phare programme, progress has been slow. There are three main reasons for this. First, the finance available has been required for the maintenance of existing roads and railways. Second, the relatively low level of traffic flows and weak purchasing power of the users of privately financed roads and railways make it difficult for these projects to break even. Third, the programme of new transport corridors adopted for Central and Eastern Europe by the European Union is very ambitious: its viability has not been justified by economic evaluation, and it far exceeds the public financing available from Western and Eastern Europe put together.

Under an initiative of the European Parliament, the second Pan-European Transport Conference was organised in Crete in April 1994. The European Union in conjunction with the European Conference of Transport Ministers, the international financing institutions, including the World Bank, EBRD, EIB and the states concerned, agreed that the principal objective was to cost the coherent development of a pan-European transport infrastructure, examine the funds available and acquire an instrument which would allow them to plan investments of common interest. Nine TENs corridors were identified, comprising road, rail and combined infrastructures, including ancillary installations such as access roads, border-crossing stations, freight and passenger terminals and traffic management (see Map Two).

Priority is to be given to those projects which meet the following criteria:

- interconnection and interoperability of international and inter-regional connections;
- maturity: the project must begin within the five-year period;
- limited impact on the environment;
- proven availability of necessary financial resources;
- an economic rate of return of at least 10%.

The 'CEE Corridors' are being coordinated by the 'G-24' Transport Working Group which brings together representatives of the OECD and Central and East European states, the European Commission and several investment banks. Headed by Ottokar Hahn, an advisor to the Commission, G-24 is stimulating investment projects and speeding up procedures for border crossings. Memoranda of understanding, designed to facilitate coordination, have been concluded for several corridors. For example, Germany, Poland, Belarus, Russia and the European Union signed a memorandum for the Berlin-Warsaw-Moscow corridor in January 1995 that provides for the extensive upgrading of rail and road facilities with the aim of cutting total travel times from 30 to 20 hours. Investment will total Ecu 5 - 10 bn.

Upgrading the nine CEE corridors to Western standards would require an estimated Ecu 30 - 45 bn for roads and another Ecu 25 - 30 bn for rail projects. The international financing

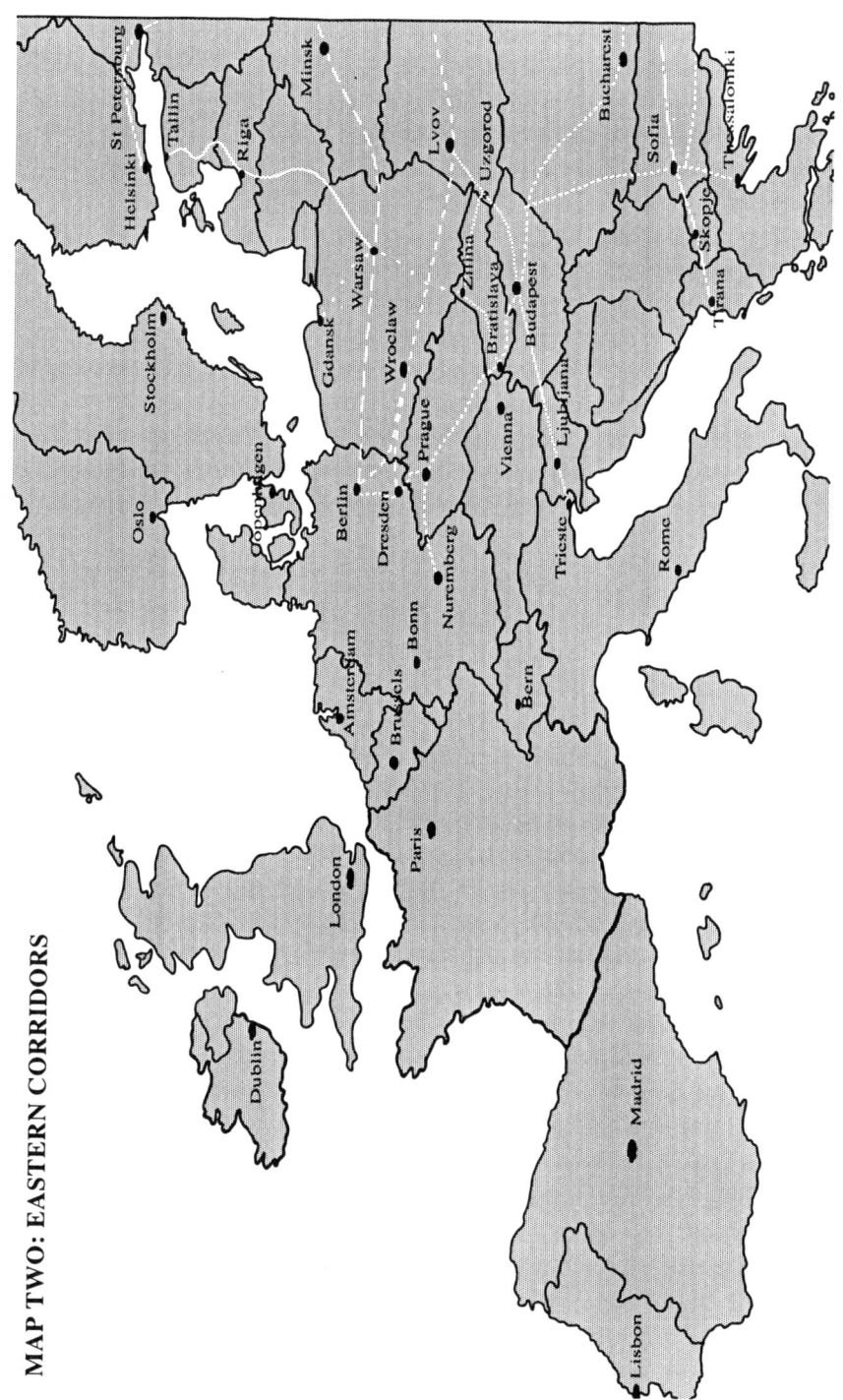

MAP TWO: EASTERN CORRIDORS

institutions had committed loans totalling Ecu 2.5 bn by mid-1995. The Phare programme will provide a further Ecu 2 bn between 1995 and 1999. Michel Gaspard of the Commission's DG VII has pointed out that on the present basis the Central and East European states will not achieve widespread modernisation. Per capita investment in Poland, for example, amounts to 20 Ecu compared to 211 Ecu in France. International bank officials estimate that under stringent financial conditions loans for the corridors could amount to Ecu 600 m per annum. Gaspard calculates that an annual Ecu 3 bn is necessary just to maintain the existing infrastructure. A real recovery would require Ecu 8 bn. [73] **Special measures are clearly needed to plug this gap, both financial and in terms of know-how.** Unrealistic expectations within national authorities, lack of reliable traffic data and legal standards, and insufficient and poorly trained staff are just some of the problems which have exacerbated a rapidly deteriorating situation.

Phare

In this situation, the Phare programme has concentrated its resources on encouraging an efficient traffic flow at a number of border bottlenecks on the main international corridors. Despite teething troubles, Phare is particularly important as it is able, unlike the banks, to provide grant-aid both for projects and feasibility studies. Most of its funding for larger projects is made available on a co-financing basis with the EIB, the EBRD and the World Bank. The Essen European Council agreed that Phare should devote a greater proportion of its funds to co-financing infrastructure developments. The limit on its annual budget for this type of spending was raised from 15% to 25%. The Council agreed that its budget should be guaranteed to provide a total of Ecu 5.5 bn in the period 1994-99. It was also agreed that Phare funding should play a greater role in planning for the TENs. Part of Phare is also dedicated to management, technical training and air traffic service, as well as the harmonisation of laws and standards.

Since the revolutions in Eastern Europe traffic has increased well beyond the capacity of the old border crossing facilities. Phare has funded improvements at the borders of Germany with the Czech Republic and Poland where border traffic increased from 12.2 m vehicles in 1989 to over 100 m in 1993 — and where delays, especially of lorries, could exceed 40 hours. The first phase of support was the transfer of know-how establishing the necessary procedures and training customs staff. The next involves the upgrading of road and railway connections, the provision of vehicle-holding areas and the rebuilding of customs points. By the end of 1995 more than 50 individual border-crossings had been modernised.

Phare also provides financial and technical assistance for the modernisation of railways. A programme, for example, has been prepared to upgrade the entire network of the Bulgarian state railways at a cost of Ecu 230 m with the EBRD, the World Bank and Phare as joint financiers. Phare's contribution is earmarked for training, safety technology, automated train guidance equipment, signal boxes and the development of a train-radio system.

EIB lending in this sector is closely coordinated with the Phare programme. EIB loans totalling Ecu 1.1 bn have been advanced for roads, railways and a programme for the modernisation of air traffic control systems. Several of these projects represent direct adjuncts to trans-European networks, in particular the Polish section of the Dresden-Kiev motorway, the Berlin -Vienna rail link and the pipeline supplying Russian oil to the European network. [74]

European Bank for Reconstruction and Development

The EBRD was established in 1991 in London to 'foster the transition towards open market-oriented economies and to promote private and entrepreneurial initiative in the Central and Eastern European countries'. [75] The Bank's mandate specifies that at least 60% of its funding should be devoted to private or privatising enterprises and 40% may be devoted to public sector infrastructure development. The private sector emphasis differentiates the EBRD from other international finance institutions such as the World Bank and the EIB.

Of necessity, half of the EBRD's transport commitments are in the state sector, compared with 38% of in other sectors. Nevertheless, the Bank is pursuing opportunities to get the private sector involved. This is particularly true for projects that support road haulage, shipping and airlines which often involve joint ventures with a Western partner and an established local company. Even in the more traditional state sector, opportunities to involve the private sector have been found. A prominent example is the M1/M15 motorway in Hungary. Other examples are airport terminals and port superstructure where the airport and port authorities, acting as landlords, lease out space for private operators to build and operate specialised terminals for concession periods. In Moldova there is an Ecu 14.8 m project to develop a river port oil-importing facility on the Danube allowing the country to meet its energy requirements more economically. This facility will be majority privately owned, the Moldovan shareholders being the largest.

In the railways sector the focus has been on restructuring to meet reduced demands. In Slovenia, the EBRD already has a railway project up and running. The main objective is to begin the process of structural adjustment towards the new role the railways will play in a market economy. To do this, it was necessary first to establish a more transparent role between the various parties. In this process EBRD has acted as an intermediary between government and the state railway. The state will subsidise loss-making passenger services in the public interest for the medium term, by way of expenditure on infrastructure. In return, the railway will increase its efficiency, partly through the rehabilitation programme financed by the project, partly through operational and service improvements, which will be monitored under the project. The successful Slovenian railway restructuring is being followed up by similar projects with the Czech and Bulgarian railways. Future projects may include private sector involvement in rolling stock leasing, maintenance work and even specialised train operations. [76]

Where there has been a willingness to pay tolls for better-than-average road services, the EBRD has been able to co-finance a public-private partnership to build and operate a motorway in Hungary. The M1/M15 motorway completes links from Budapest to Bratislava and Vienna. Public service concessions are regulated in Hungary under a 1991 law which makes it possible for a private consortium to finance, build and operate motorways owned by the state. The concession contract can extend up to 35 years. Somewhat unusually, this contract is regulated by Hungarian law, whereas the construction, maintenance and operation, loan agreements are governed by English law as a result of Western construction partners being supported by British financial and American legal advisers chosen by the government. In some ways this project was a special case as the main construction contract provides for a missing final link of 43 km in a continuous motorway between Vienna and Budapest with a 15 km branch towards Bratislava. As around half of the traffic comprises Western-registered vehicles with relatively high time values, there is a corresponding willingness to pay tolls. The government is contributing the land for the motorway construction and expects to receive net payments from the concessionaire in the form of fees and profit sharing (15%). However, the government is not providing equity, loans or guarantees to the project. The EBRD took the lead in putting together an Ecu 280 m debt-financing package with an innovative mix of instruments, including long-term local currency loans and bonds designed to minimise currency risk and stretch maturities to over 15 years.

Lessons to be learnt from the M1/M15 motorway project have been summed up as follows:

- sophisticated project financing can be achieved in Central and Eastern Europe;
- it is possible to fund projects locally — in this case up between 30 - 50%;
- the revenue stream was enhanced by the high proportion of foreign users;
- the project is cross-subsidised: for the toll fee, motorists will be allowed to drive over the existing 128 km motorway, as well as the newly constructed 43 km. [77]

The M1/M15 motorway project's main element, the M1 section, was completed on time and within budget, and was opened early in January 1996. The observed revenue stream falls below forecast but does not yet jeopardise limited recourse financing.

The EBRD has taken the lead in the financing of the M5 toll motorway, running from Budapest to Hungary's southern border. Financial closing was achieved in December 1995. Construction started in March 1996. The government is providing as its in-kind contribution the first phase of the existing 27 km motorway and 30 km half-motorway, as well as land to build the second carriageway along the latter, and a further 40 km new motorway section. An Ecu-linked stand-by type operational subsidy from the Road Fund under sovereign guarantee is also available during the first 6.5 years of operation. The lending syndicate comprises 16 international and 4 local banks. The EBRD will provide hard currency loans, offer guarantees to the banks for the final repayment of their loan as well as lending one-third of total financing of Ecu 204 m.

However, a motorway link east of Budapest (M3/M30), which was also to have been built as a public/private partnership, is now to be implemented as a public sector toll motorway project, using EIB and other loans under sovereign guarantee. The Hungarian government rejected the concession proposal of GTMI as too expensive compared with the public finance alternative. [78]

Lessons for the West

There are lessons for the European Union in the ways that Central and Eastern Europe is tackling its formidable infrastructure problems. First, there is the corridor approach. The TENs corridor is based on an intermodal analysis that examines the contribution which can be made by road, rail and combined transport as well as ancillary installations such as access roads, border crossings, service stations and freight and passenger terminals. Second, a priority must be given to getting the conditions right for the participation of the private sector in the development and the operation of the corridor. In order to promote public-private partnerships, the private sector has to be involved at the earliest stages of planning the implementation and operation of the networks. Furthermore, all those involved must agree a common set of technical norms necessary to secure interoperability of all sections of a corridor.

A massive modal shift is taking place in Central and Eastern Europe. Forecasts for the period 1990 - 2005 predict that the share of road traffic between East and West will double while rail traffic will decline sharply. The total traffic volume is estimated to grow from 73 m to 170 m tonnes in the same period. Faced with a collapsing railways sector and the threat of congested roads, the Central and East European states are trying hard to keep some freight on rail. For example, a task force has been established for the Berlin - Black Sea corridor comprising all relevant private and public stakeholders to formalise various ideas for promoting combined transport — including a supranational wagon pool, a regional terminal network, introduction of telematic systems and a corridor development company.

Catalytic role

Another prominent feature in the development of public-private partnerships in Central and Eastern Europe is the role of the international institutions. Phare, the EBRD, World Bank and the

EIB play a much wider role than the provision of finance. They are involved in the establishment of administrative and regulatory frameworks, and in the selective unbundling of the old public monopolies in order to provide room for the private sector. The benefits can be seen in telecoms. International banks can play a more overtly catalytic role in Central and Eastern Europe because their financial umbrella is a *sine qua non* of Western investment capital. As Chancellor Vranitzky has pointed out, the EBRD and the World Bank are increasingly concentrating on covering for the quasi-political risk without which privately financed infrastructure will not take place. **New structures and instruments have been successfully created to match sovereign and private borrowing, and, as part of the pre-accession strategy, to prepare the applicant states for accession to the Union.** [79]

There is still considerable scope, however, for the pooling of investments in order to reduce the transaction costs of private investment in Eastern Europe and, at the same time, to increase the size of the pool. The EBRD has pioneered a system of framework agreements that streamlines the financing of a number of small and medium-sized investments. The Nordic Environmental Finance Corporation (NEFCO) has demonstrated that it is possible to set up multilateral risk capital funds for investing in private sector joint ventures in Central and Eastern Europe. NEFCO's policy is to seek viable projects which enable an investment to be implemented and disinvested as soon as possible, through allowing a buy-out on predetermined conditions, but with an upside for NEFCO. With its public budgetary funding, NEFCO is able to assume higher and longer-term risks than a private venture capital fund would be able to manage. [80]

6 | A European Infrastructure Agency

In this final chapter we sum up our findings and make proposals for the setting up of a European Union Infrastructure Agency.

European Councils since Essen in 1994 have repeatedly stressed the contribution that the TENs could make to the single market and to the competitiveness of the European economy. The Ciampi group set up to advise the EU on competitiveness has drawn attention to the benefits that could be gained from encouraging private enterprise into the financing and running of public utilities. President Santer's European Pact of Confidence for Employment presented to the Florence Summit in June 1996 emphasised that the time had come to remove the last remaining obstacles, whether financial or technical, to the construction of the TENs. It said that:

> 'The Commission has already made proposals for providing the additional financing expected from the community (Ecu 1.2 bn asked for extra TENs budget lines since Essen). A decision must be taken on these proposals once and for all.

The Commission also proposed that with the EIB a fresh look be taken at the Bank's role in financing the networks. Lastly, the Commission is very firmly in favour of a considerable increase in the number of public sector-private sector partnerships. It will set up a high-level group to press for progress on this'. [81]

In the event, Germany and the UK blocked the extra financing of Ecu 1.2 bn needed to accelerate the TENs projects on the grounds that there were technical difficulties in launching the networks, including the lack of compatible standards, as well as insufficient confidence among private investors.

The Christophersen Group in its report to the Essen European Council drew attention to the fact that the difficulties facing the priority projects can be traced back to one or a combination of the following features of the projects:

- trans-frontier nature;
- large scale;
- high risk;
- limited financial viability. [82]

The TENs programme is now stalled within the inter-institutional procedures of the Union. The European Council at Florence in June 1996 did not endorse President Santer's proposals for additional EU grant financing for TENs. Instead it decided to set up another High-Level Group to press for further progress.

It is difficult to see how yet another expert group will be able to reconcile the problems involved without both a secretariat and some increase in the TENs budget lines. In our view, **a new Task Force should be set up immediately by the Commission to bring together the existing expertise which is at present dispersed among the various DGs and other European institutions. It would have the powers of a directorate-general. Its remit, which would be time-limited, should be to expedite the TENs programme and to facilitate the contributions of public-private partnerships.**

The Task Force and the High-Level Group should together prepare the ground for the establishment of a new independent body — a *European Infrastructure Agency* — with powers to evaluate TENs projects from a European dimension. Such an agency would be able to draw on the expertise of the European Commission, the European Investment Bank, the European Investment Fund, the EBRD and member governments and corporations as well as the private sector. The Australian Infrastructure Advisory Committee is analogous.

In the longer term there is a strong case for endowing the Agency with limited executive powers. It would then be able to take a lead in arbitrating on the distribution of public sector contributions and benefits arising from transnational schemes. Such an enhanced authority would be able, in conjunction with the European Investment Fund, to encourage and foster the replacement of grant finance with public equity thereby creating the basis for more private participation and the eventual development of a European Infrastructure Fund.

Network effect

The idea of a 'high-speed network project authority' was strongly endorsed by the High-Level Group set up to consider the European high-speed train network. This group, which included representatives of the European Commission, the EIB, member state governments and the Union of European Railway Industries, proposed that the authority's remit would consist, *inter alia,* of:

- giving new impetus to network construction;
- devising procedures for retroceding revenue between different infrastructure operations;
- mobilising EU aid, particularly for the key links.

'As the issues at stake transcend purely national interests, resolute Community action is needed to ensure network interoperability. The concept of an authority, the characteristics of which still need clarifying, would seem to constitute the best approach for meeting identified needs.' [83]

European transport ministers have also emphasised the absence of a Europe-wide consideration of infrastructure needs. They have drawn attention to the likelihood that the annual costs of traffic congestion could grow from an estimated Ecu 4 billion in 1990 to Ecu 14 billion by 2010. International traffic is the most dynamic element, and the European Commission expects an increase in international road freight by as much as 156% by 2020. The OECD suggests that measures so far adopted have proved inadequate because they have consisted of essentially national and highly differentiated approaches based on scenarios and planning processes that differ considerably from one country to another. [84]

In particular, traffic assessment and forecasting studies are carried out on an individual basis by means of methods that are not systematically comparable. Freight and passenger traffic statistics are seldom shown on a comparative basis, making a proper

TABLE NINE

Comparison of Key Features of Railway Evaluation Methods in EU Member States

	BASIC EVALUATION FEATURES	SOCIO-ECONOMIC MEASURES	DISCOUNT RATE	MAJOR EXCLUSIONS	SPECIAL FEATURES
BELGIUM	Financial and socio-economic appraisals in context of rail policy STAR 21	NPV, IRR	6% and various other	International user benefits excluded	HSR in Belgium is essentially an international service. Balance between regions is very important
DENMARK	Socio-economic appraisal in context of strategic decisions			Corridor impact appraisals may exclude some network effects	Strategic importance of linking national and international sea crossings of paramount importance
FRANCE	Parallel financial and socio-economic evaluations	NPV, IRR	8%	Diversions from road not appraised separately	Socio-economic evaluation monitors transfers to air transport carriers
GERMANY	Integrated socio-economic evaluation of all surface transport	NPV benefit/cost ratio	3%	Newly generated traffic not appraised	Low discount rate and capital rationing gives bias to long-term projects
GREECE	New transport investments including rail, normally appraised by socio-economic cost-benefit studies	NPV, IRR	6% and other values		Financial problems constrain new investment
IRELAND	New transport investments, including rail, normally appraised by socio-economic cost-benefit studies	NPV	5%		Financial problems have constrained new investments
ITALY	Financial and socio-economic evaluations in context of wish to introduce private finance	NPV	8%		Socio-economic criteria and the need to satisfy new conditions of private participation will need to be reconciled
NETHERLANDS	Socio-economic evaluations in context of goal-oriented transport policy. Rail 21 is development plan for national railways	NPV & IRR for international proposals	Various	International user benefits excluded	HSR in the Netherlands would be mainly an international service. Interest in environment is very high
PORTUGAL	Rail transport evaluated by cost-benefit appraisal in context of official rail investment programme				Need to balance national needs with desire to link with European network. Non-standard gauge a feature of Portuguese railways
SPAIN	Official manual for financial, economic and social appraisals of rail investments	NPV	6%		Environmental impacts included in 'social' evaluation. Non-standard gauge a feature of Spanish railways
UNITED KINGDOM	Basically financial appraisal with partial cost-benefit appraisals	NPV	8%	Many user benefits excluded	Comprehensive impact appraisals including some assessment of land-use impacts. User benefits excluded except for subsidized services. Interest in environment very strong

Note: All member states carry out environmental impact assessments and internalize some environmental impacts by use of remedial measures.

Source: *High-Level Group for High-Speed Rail Network Report from Financial Sub-Group* Vii:93.

assessment of networks impossible. Furthermore, as can be seen in Table Nine, EU member states have different rules for calculating cost-benefit analysis as well as for evaluating the investment choices. As the High-Level Group has admitted: 'It is therefore hard to compare the results of the various methods, particularly since the techniques of forecasting, modelling and estimating impacts are in a continual state of flux and differ from one project and from one country to another'. [85] It is particularly frustrating that a number of member states do not include the benefits to foreign passengers in their evaluations. [86] This failure can make a major TEN, such as PBKAL, succeed or otherwise in meeting the desired internationally acceptable rate of return in both economic and financial terms. It is a major obstacle to planning network development, and impairs decisions about the level of any EU funding.

A principal function of a TENs Agency, therefore, would be to evaluate projects from an overall European point of view and to harmonise the assessment of potential projects. **What is clearly required is a common methodology for network Europe.**

The European Infrastructure Agency should also be responsible for establishing a common mechanism for transferring costs and benefits between old and new sections of the network. In principle the last operator to slot into the network always has the advantage of exploiting both existing and untapped traffic demand. A situation whereby each promoter waits for the other to make the first move, as has happened with motorway networks, could occur here too. As the High-Level Group pointed out: 'So if the network, and particularly the key links are to be completed quickly, it is essential to establish the principle of redistribution, which requires the beneficiaries to make a certain contribution towards the new sections, in conjunction with the appropriate compensatory mechanisms'. [87]

Paying for public-private partnerships

The complex financing packages, involving both the public and private sector, which are necessary to fund the priority TENs call for management and not just legislative directives and regulations. In the absence of a specialised TENs agency with the ability to reconcile the different partners' interests and organise compromises, the construction of the network links could take too long for the potential technical, economic and political advantages to be secured. A characteristic of many TENs is the long period that elapses before there is any revenue to meet financing charges. In particular, the initial planning phase typically requires public sector budgetary grant financing and/or high risk venture capital equity financing.

At present grants or their equivalent are extensively used by the EU and member states to cover 'equity gaps' in major transport projects. The problem is that it is difficult to avoid excessive subsidising as grant decisions have to be made early in the project cycle at a time when it is difficult to quantify project risks accurately. The analysis of 'The Role of Private Finance' in Chapter Four suggests that public equity or quasi-equity could be more desirable especially if no dividend was paid or expected until revenue flows were realised. Grant financing has been preferred as it enables the public sector to conduct a hands-off management in relation to private partners. Another reason is that the public sector, whether the European Commission or national Treasuries, have little expertise in managing equity stakes and, consequently, are reluctant to be involved especially if they are afterwards held to account by public auditors on an *ex post facto* basis.

These difficulties need not apply to an independent TENs agency which would be able to encourage much needed equity investment in projects both by the public and private sectors. Venture capital would be more forthcoming from operating consortia and specialist boutiques if there were a growth of institutionally backed infrastructure funds similar to the billion dollar pools that have been encouraged by the World Bank in Asia and Latin America. In due course **it is possible to envisage a revolving European Infrastructure Fund that would take equity and quasi-equity stakes in TENs projects.**

The role of the TENs Agency would include acting as a catalyst in seeking out and refining new financing instruments. In doing so, it would be working alongside the European Union's structural and cohesion funds, the European Investment Bank, the European Investment Fund, and the private sector banks. So far the EIF has a very limited role in the provision of equity and quasi-equity for TENs projects. The establishment of a TENs Agency could act as a powerful lobby drawing attention to the need for more public and private equity in its published evaluation of the required complex financing packages.

Structural and cohesion funds

The TENs Agency may be able to encourage the focusing of the EU's structural and cohesion funds on TENs. The cohesion fund, in particular, has already made a substantial contribution to transport infrastructure. During 1993-95 the fund channelled Ecu 3.4 billion into TENs in Greece, Ireland, Spain and Portugal. Major road projects have been the main beneficiaries, but significant assistance has been allocated to high-speed and conventional railway systems. Four of the eleven mature projects are located in the cohesion countries. As the discipline of Stage Three of EMU begins to bite, there is a strong case for insisting that the cohesion fund increasingly generates projects of better value for money through the greater participation of private sector entrepreneurship via public-private partnerships.

The structural funds do not usually co-finance infrastructure with the private sector, although the funding provided between 1993-95 for TENs projects has not been insignificant with Ecu 1 bn provided for transport, Ecu 763 m for energy and Ecu 295 m for telecom TENs. While preparing the EU support frameworks, the Commission has given a high priority to trans-European networks. Projects include the introduction of natural gas in Greece and Portugal and the interconnection of the Greek and Italian electricity networks.

Best value for money

A TENs Agency would be able to take on board the best practice gained in public-private partnerships in the member states. The French system of granting agreements to concessionaires before all the details are known has proved less rigid than the British approach to PFI. In particular France has adopted a flexible approach to the sharing of any surplus profits carried over and above the expected rate of return. They provide partial self-finance for subsequent projects. This financial transfer mechanism underpins the French development of roads, water supply and public transport. The acceptance of public equity or quasi-equity in place of grants would go a long way to developing a European system that would also avoid some of the problems that have arisen in the UK. [88] Public equity would make TENs more attractive in terms of tax depreciation to the private partners. At present capital and revenue subsidies are normally taxable, but investment in TENs would be much more attractive if capital subsidies were regarded as quasi-equity and not as reducing the cost of assets for tax depreciation purposes.

The contribution that the UK Private Finance Initiative could make to the work of a TENs Agency could also be considerable. There can be little doubt that as experience of PFI grows, officials will become more skilled at evaluating projects. One of the best prospects among the TENs so far is the English Channel Tunnel Rail Link — part of the PFI described in Chapter Three. In particular, the UK has been in the forefront of developing evaluation methods which are innovative and transparent. The PFI delivers better value for money by producing a different solution from the traditional European public sector approach. Once again **a TENs Agency could play an important role in the EU by demonstrating how privately-led delivery savings can outweigh the cheaper cost of public sector borrowing**.

At the same time it may be necessary to show that the European tax payer is getting value for money. There has been criticism of the complex public sector comparators that are used by the British Treasury. [89] On the other hand, **work done by leading actuaries in the UK has demonstrated that the techniques of risk analysis, risk management and project finance can provide a relevant evaluation**. [90]

Transnational operating companies

A TENs Agency would be able to act as a catalyst in the development of transnational, risk-sharing operating companies. The traditional inclination of construction companies has been to work on contract to a project promoter (often the public sector) and take as little of the risk relating to the project as possible. It would clearly be very helpful in the development of TENs if there could be an early completion of the EU's legislative framework for European companies.

In Central and Eastern Europe, the EBRD has been encouraging the development of public-private partnerships through the unbundling of state monopolies in telecoms and energy. A TENs Agency would be able to act as a catalyst for the unbundling and the development of user charges and thereby encourage the creation of transnational operating companies similar to the London and Continental Railway which will be the owner/operator of the English

Channel Tunnel Rail Link. In any case, the European Commission would find it difficult to carry out a comparable role — even assuming it had the right personnel to carry out this function.

Framework agreements

One of the basic reasons for protracted negotiations and frustration over TENs is misunderstanding about the degree of perceived and real risks in a particular project. On the other hand risk sharing is fundamental to public-private partnerships. Risk should be allocated to the partner best able to manage it. The aim is not to encourage risk transfer for its own sake but to achieve the most cost-effective solution. **An autonomous TENs Agency could provide a framework for the exchange of information, the settlement of conflicts and the pooling of non-commercial risk**. There is growing evidence that the lack of credible statutory and regulatory frameworks is a critical barrier to attracting and sustaining private investment. Under a case-by-case approach, arguing that each project is unique, sponsors of individual ventures can try to negotiate mitigation of each and every risk by relevant state authorities. The World Bank has encouraged the development of 'templates' for each major sector, clarifying ahead of competition who will bear the risks and how. This way, all potential participants are treated equally and know the rules of the game before submitting proposals.[91]

A TENs Agency would also be able to encourage innovative arrangements such as the framework agreements pioneered by the EBRD. These provide a streamlining of the financing of small and medium-sized investments for the public-private provision of municipal services. The EBRD and the World Bank are increasingly providing cover for the quasi-political risk without which privately financed infrastructure will not place in Central and Eastern Europe. Once again, an autonomous EU Agency which built up a reputation for establishing best practice in evaluating projects would be well placed to encourage the development of such risk-sharing instruments.

Relationship with EU institutions

In the Treaty, the powers of the EU institutions are defined ambiguously. Article 3(n) lays down that one of the activities of the Community shall be 'encouragement for the establishment and development of trans-European networks'. Article 129b says that the Community 'shall contribute to the establishment and development of trans-European networks in the areas of transport, telecommunications and energy infrastructures'. EC action shall 'aim at promoting the interconnection and interoperability of national networks as well as access to such networks', particularly linking the core with the periphery.

In all this — more fully explained in Chapter One — the European Commission plays a key role in initiating projects and in building the political package. It is likely to be less accomplished at creating the right financial conditions to facilitate the projects, and has faced stiff political opposition from certain member state governments in creating an independent financial capacity on the line of 'Union bonds'. The Commission's deployment of the R&D, know-how, structural and cohesion funds is uncoordinated and unfocused. And the Commission has a very poor track record in managing contracts.

Internally, the Commission faces an enormous struggle to coordinate adequately the stance of the various Directorates-General involved: DG II (Economics & Finance), DG III (Industry), DG IV (Competition), DG VII (Transport), DG XI (Environment), DG XIII (Telecoms & Information Market), DG XV (Single Market), DG XVI (Regional Policy), DG XVII (Energy), DG XVIII (Credit), DG XIX (Budget) and DG XX (Financial Control). In the case of Central and Eastern Europe, DG I (External Economic Relations) is also heavily implicated.

In matters of funding, the Commission cooperates with the EIB and, increasingly with regard to Central Europe, the EBRD. Although all three organisations bring their own expertise to the problem, there is no single methodology for the assessment of financial risk. Nor is there yet one accepted definition of the criteria of universal service obligation, and the concept of 'public utility' differs markedly between member states.

Clearly, in these circumstances the public authorities cannot rely wholly on the private sector to appraise the feasibility of a project; nor can the private sector be expected to take the word of the public authorities on trust. Both require the reassurance of the other, and the ability of the EU to build and operate its TENs depends entirely on the quality of the partnership and the sharing of risk.

On a political level, the Commission has to steer any package through the Council and Parliament. It also has to deal directly with the national governments of the member states concerned and with regional and local authorities, especially in the federated states of Austria, Belgium and Germany. Trans-alpine infrastructure projects involve crucially the non-member state of federal Switzerland.

One major problem in this as in other matters is the fact that the Commission's resources are inadequate to the size of its tasks. As things stand at present the Commission's DGs and, albeit to a less extent, the EIB, are forced to rely on the information they receive from the private sector. **We are proposing the establishment of a new specialised EU Agency for TENs whose functions would include the provision of independent advice to the Commission, Council and Parliament**. Such an arms' length agency would be especially valuable in guiding the competition and state aids policy for TENs, where the Commission is at present legislator, watchdog, investigator, judge and enforcement authority.

There is already an argument being advanced at the IGC for drawing powers away from DG IV and up from the monopolies, mergers and fair trading authorities of the member states to create a new supranational competition tribunal. Even if such a radical approach to competition policy were not adopted, the creation of a new advisory agency for TENs, drawing technical and legal expertise from the services of the Commission and member state governments but also on secondment from the private sector, is an attractive idea. The EU Infrastructure Agency would need a Treaty base, would have to report annually to the Commission, Council and European Parliament, and be made subject to the judicial review of the Court of Justice. **We recommend that this proposal is included on the agenda of the current Intergovernmental Conference** which is already examining the parallel question of an EU competition agency. The Commission could retain its wide administrative discretion, but how and when it were to exercise it, and how it reached its decisions, would be both more transparent and better founded.

Paradigm for the twenty-first century

At the member state level, especially in France and the United Kingdom, public-private partnerships are beginning to demonstrate how they can deliver more efficiently the provision of infrastructure. It is true that the cheaper credit available to governments needs to be weighed against possible inefficiencies that arise when financial discipline is relaxed as a result of government sponsorship. The World Bank has pointed out that cost overruns and time delays are common in purely public sector provision leading to cumulative cost increases that can easily cancel out any interest rate advantage enjoyed by government, whereas public-private partnerships have to deliver on time.

Consumers would undoubtedly benefit if it were possible to combine low interest rates and efficient provision. But even credit-worthy governments cannot borrow unlimited amounts at low cost. An ageing population is going to put enormous pressures on public finances into the twenty-first century. A consequence will be that **the downward trend on capital expenditure on infrastructure may well continue unless the private sector plays a more important role**. It is to be remembered that in the nineteenth century the private sector was the leading provider. In the twentieth century the public sector took over the role. In the twenty-first century, guided by a European Union Infrastructure Agency, public-private partnerships could become the central provider of the infrastructure of network Europe.

NOTES

[1] *Missing Links,* European Round Table of Industrialists, Brussels, 1984.

[2] Article 189b procedure.

[3] Article 189c procedure.

[4] *Growth, Competitiveness, Employment. The challenges and ways forward into the 21st Century,* EC *Bulletin,* Supplement 6/93.

[5] *Enhancing European Competitiveness,* Second (Ciampi) Report to the President of the European Commission and Heads of State and Government, Luxembourg, December 1995.

[6] *Trans-European Networks,* Group of Personal Representatives of the Heads of State and Governments (Christophersen Report), Report to Essen European Council, 1995.

[7] Christophersen Report, op. cit., p. 12.

[8] Essen European Council, Presidency Conclusions, EC *Bulletin,* 12/1994, para. I.6.

[9] EC *Bulletin,* 'Europe and the global information society: recommendation to the European Council', (Bangemann Report), Supplement 2/1994.

[10] See Federal Trust Report (Andrew Adonis, rapporteur), *Network Europe and the Information Society,* London, Federal Trust, 1995.

[11] *Progress on TENs,* Report to Madrid European Council, COM (95) 571, para. 2.2.

[12] See EC *Bulletin,* 5/1996, para. 2.2.1.

[13] Wolfgang Hager, 'Industry needs and can deliver efficient infrastructures' in *The Provision of Infrasructure; the Role of the Private Sector,* Proceedings, EIB Forum, 1995.

[14] Claude Martinand (ed.), *Private Financing of Public Infrastructure,* Paris, DAEI, 1995, p. 13.

[15] Dominique Lorrain, *Urban Services, The Market and Politics,* in Martinand, op.cit., p. 47.

[16] See Jean-François Poupinel, *The French Highway System,* in Martinand, op.cit.

[17] *Private Opportunity, Public Benefit: Progressing the Private Finance Initiative,* HM Treasury, 1995, para. 2.1.

[18] David Willetts, *The Opportunities for Private Funding in the NHS,* London, Social Market Foundation, 1993, p. 5.

[19] T. Barnekov, R. Boyle, D. Rich, *Privatism and Urban Policy in Britain and the USA*, London, OUP, 1990, p. 191.

[20] Brian Robson, *Assessing the Impact of Urban Policy*, London, HMSO, 1994, para. 5.12.

[21] *Private Opportunity, Public Benefit,* op. cit., para. 3.35.

[22] House of Commons, *The Private Finance Initiative,* Treasury and Civil Service Select Committee, Sixth Report, HC 146, London, HMSO, 1995-96, p. XV.

[23] *Private Opportunity, Public Benefit,* op. cit., para 2.7.

[24] *Private Opportunity, Public Benefit,* op. cit., para. 4.64.

[25] *Private Finance Initiative,* op. cit., p. XIV.

[26] *Private Finance Initiative,* op. cit., p. 53.

[27] The Treasury notes that the Net Present Value of the preferred DCMF contracts was 'significantly less' than the public sector base case. (*Private Opportunity, Public Benefit,* op. cit., para.4.40).

[28] *Private Finance Initiative,* op. cit., p. XVI.

[29] Coopers and Lybrand, Memorandum of Evidence to the House of Commons, *Private Finance Initiative,* op. cit., p. 29.

[30] Coopers and Lybrand, op. cit., p. 33.

[31] Coopers and Lybrand, op. cit., p. 34.

[32] See Harry Cowie, *The Phoenix Partnership - Urban Regeneration for the 21st Century,* London, National Council of Building Material Producers, 1985.

[33] See S. Fosler & R.A. Berger (eds), *Public-Private Partnership in American Cities,* Lexington, Mass., Lexington Books, 1982.

[34] US Department of Housing and Urban Development, *Public-Private Partnerships,* Washington, 1984.

[35] Robin Hambleton, 'Not on EZ Street', *The Guardian*, Society Supplement, 15 November 1995.

[36] *Assessing the Impact of Urban Policy, op.cit.,* para. 5.13, p.50.

[37] *World Bank Development Report 1994, Infrastructure for Development,* World Bank, OUP, 1994; *Private Infrastructure Task Force report,* Economic Planning Advisory Commission, Canberra, 1995.

[38] World Bank, *Infrastructure Development in East Asia and Pacific - Towards a New Public-Private Partnership,* Washington, 1995, p. 1.

[39] Pang Chung Min, *Asian Experience in Raising Private Finance*, DG II Workshop on Private Sector Finance of Community Interest Projects, European Commission, Brussels, June 1994.

[40] Ian Jones, Hadi Zamani, Rebecca Reehal, *Financing Models for New Transport Infrastructure: Report to DGVII European Commission*, London, National Economic Research Associates (NERA), 1996, p. 24.

[41] Directive 91/440.

[42] Neil Kinnock, *The Trans-European Transport Network and the Role of Public/Private Partnerships,* Address to IBC Conference, Brussels, 27 February 1996. See also *Making it Happen; Building and Financing TENs,* European Centre for Infrastructure Studies, ECIS Report, December 1994.

[43] Robert Boas, *Financing Transport Infrastructure Projects,* Address to IBC Conference, Brussels, 27 February 1996.

[44] Sir Alastair Morton, Memorandum of Evidence to the House of Commons, *Private Finance Initiative,* op. cit., p. 119.

[45] European Commission, *Financing of Transport TENs by Public Private Partnerships*, Report II/266, 1995. However, France, Italy, Germany and Sweden now allow regional subsidies for railway infrastructure.

[46] *Cofiroute*, Annual Report 1993.

[47] *Private Finance Initiative,* op. cit., p. XV.

[48] See Poupinel in Martinand, op. cit.

[49] Andrew Adonis, 'Private Finance Initiative Attacked', *Financial Times,* 26 March 1996.

[50] Coopers & Lybrand, op. cit., p. 32.

[51] See *Making it Happen,* op. cit.

[52] *Progress on TENs*, Report to Madrid European Council, op. cit., Annex 7.

[53] See Peter Laurson, *Private sector financing for trans-European transport networks - actions to overcome constraints,* DG II European Commission, November 1994.

[54] Laurson, op. cit.

[55] *Financing Models for a New Transport Infrastructure,* op. cit., p. 33.

[56] *Making it happen*, op. cit.

[57] Ranjit Mathrani, *Financial Engineering of large infrastructure projects*, IBC conference, Brussels, 27 February 1996.

[58] See *Private Infrastructure Task Force ,* Canberra, op, cit.

[59] André Taylor, 'PFI fund draws two big backers', *Financial Times,* 30 January 1996.

[60] Sir Brian Unwin, EIB Forum, Amsterdam, 18-19 May 1995.

[61] David McGlue, *Trans-European Networks: How can EIF Contribute?*, IBC Conference 'Public-Private Partnerships in Trans-European Networks', Brussels, 27 February 1996.

[62] See Federal Trust Report (Dick Taverne, rapporteur), *Towards an Integrated European Capital Market*, London, Federal Trust, 1993.

[63] See Federal Trust Report (Dick Taverne, rapporteur), *The Pension Time Bomb in Europe*, London, Federal Trust, 1995.

[64] Unwin, Amsterdam, op. cit.

[65] European Commission, *Financing of Transport TENs*, op. cit.

[66] See Roy Rana, *Lost and Found: The Community Component of Economic Return on the Investment in PBKAL*, ECIS Report, November 1995.

[67] European Commission, *Financing of Transport TENs*, op. cit.

[68] European Commission, *Towards Fair and Efficient Pricing in Transport: Policy Options for Internalising the External Costs of Transport in the European Union*, December 1995.

[69] Koen De Ryck, *European Pension Funds: Their impact on European Capital Markets and Competitiveness*, Brussels, European Federation for Retirement Provision, 1996, p. 1.

[70] EBRD, *1995 Annual Report*, London, European Bank for Reconstruction and Development, 1995, p. 23.

[71] Thierry Baudon, *Overview of Private Infrastructure Financing Prospects in Central & Eastern Europe*, EBRD Paper, London.

[72] EBRD, *1995 Annual Report*, op. cit.

[73] *New Transport Corridors for Central and Eastern Europe*, ECIS Newsletter No.4, Rotterdam, March 1995.

[74] *The EIB and trans-European Networks*, EIB Information No.86, Luxembourg, November 1995.

[75] EBRD, *1995 Annual Report*, op. cit.

[76] See Graham Smith, *EBRD Involvement in Modernising Transport Systems in Central and Eastern Europe*, London, EBRD, 1994.

[77] EBRD, *1995 Annual Report*, op. cit.

[78] *Hungary cancels motorway concession*, ECIS Newsletter No. 6, Rotterdam, November 1995.

[79] For a full discussion, see Federal Trust Papers No. 5, *Enlarging the Union*, London, Federal Trust, 1996.

[80] Laurson, *Private Sector Financing*, op. cit.

[81] Memorandum from President Santer, *European Pact of Confidence for Employment*, June 1996.

[82] *Christophersen Report*, op. cit.

[83] High-Level Group for High-Speed Rail Network *High-Speed Europe,* Brussels, February 1995, p. 163.

[84] European Conference of Transport Ministers, *European Transport, Trends and Infrastructure Needs,* Paris, OECD, 1995.

[85] High-Level Group for High-Speed Rail Network, *Report from the Financial Sub-Group,* VII 1993.

[86] Rana, *Lost and Found,* op. cit.

[87] High Level Group, *Report from the Financial Sub-Group,* op. cit.

[88] *Private Finance Initiative,* op. cit. p. XV.

[89] CBI, *Private Skills in Public Service - Tuning the PFI,* London, July 1996.

[90] Chris Lewin, 'Financial Appraisal of Project Risk', *PFI Journal,* Vol. 1, Issue 1, January 1996.

[91] World Bank, *Infrastructure Development in East Asia and Pacific - Towards a new public-private partnership,* Washington DC, 1995.